TRANSITION FROM OPERATION TO DECOMMISSIONING OF NUCLEAR INSTALLATIONS

The following States are Members of the International Atomic Energy Agency:

AFGHANISTAN
ALBANIA
ALGERIA
ANGOLA
ARGENTINA
ARMENIA
AUSTRALIA
AUSTRIA
AZERBAIJAN
BANGLADESH
BELARUS
BELGIUM
BENIN
BOLIVIA
BOSNIA AND HERZEGOVINA
BOTSWANA
BRAZIL
BULGARIA
BURKINA FASO
CAMEROON
CANADA
CENTRAL AFRICAN REPUBLIC
CHILE
CHINA
COLOMBIA
COSTA RICA
CÔTE D'IVOIRE
CROATIA
CUBA
CYPRUS
CZECH REPUBLIC
DEMOCRATIC REPUBLIC OF THE CONGO
DENMARK
DOMINICAN REPUBLIC
ECUADOR
EGYPT
EL SALVADOR
ERITREA
ESTONIA
ETHIOPIA
FINLAND
FRANCE
GABON
GEORGIA
GERMANY
GHANA
GREECE
GUATEMALA
HAITI
HOLY SEE
HONDURAS
HUNGARY
ICELAND
INDIA
INDONESIA
IRAN, ISLAMIC REPUBLIC OF
IRAQ
IRELAND
ISRAEL
ITALY
JAMAICA
JAPAN
JORDAN
KAZAKHSTAN
KENYA
KOREA, REPUBLIC OF
KUWAIT
KYRGYZSTAN
LATVIA
LEBANON
LIBERIA
LIBYAN ARAB JAMAHIRIYA
LIECHTENSTEIN
LITHUANIA
LUXEMBOURG
MADAGASCAR
MALAYSIA
MALI
MALTA
MARSHALL ISLANDS
MAURITIUS
MEXICO
MONACO
MONGOLIA
MOROCCO
MYANMAR
NAMIBIA
NETHERLANDS
NEW ZEALAND
NICARAGUA
NIGER
NIGERIA
NORWAY
PAKISTAN
PANAMA
PARAGUAY
PERU
PHILIPPINES
POLAND
PORTUGAL
QATAR
REPUBLIC OF MOLDOVA
ROMANIA
RUSSIAN FEDERATION
SAUDI ARABIA
SENEGAL
SERBIA AND MONTENEGRO
SEYCHELLES
SIERRA LEONE
SINGAPORE
SLOVAKIA
SLOVENIA
SOUTH AFRICA
SPAIN
SRI LANKA
SUDAN
SWEDEN
SWITZERLAND
SYRIAN ARAB REPUBLIC
TAJIKISTAN
THAILAND
THE FORMER YUGOSLAV REPUBLIC OF MACEDONIA
TUNISIA
TURKEY
UGANDA
UKRAINE
UNITED ARAB EMIRATES
UNITED KINGDOM OF GREAT BRITAIN AND NORTHERN IRELAND
UNITED REPUBLIC OF TANZANIA
UNITED STATES OF AMERICA
URUGUAY
UZBEKISTAN
VENEZUELA
VIETNAM
YEMEN
ZAMBIA
ZIMBABWE

The Agency's Statute was approved on 23 October 1956 by the Conference on the Statute of the IAEA held at United Nations Headquarters, New York; it entered into force on 29 July 1957. The Headquarters of the Agency are situated in Vienna. Its principal objective is "to accelerate and enlarge the contribution of atomic energy to peace, health and prosperity throughout the world".

Printed by the IAEA in Austria

April 2004
STI/DOC/010/420

TECHNICAL REPORTS SERIES No. 420

TRANSITION FROM OPERATION TO DECOMMISSIONING OF NUCLEAR INSTALLATIONS

INTERNATIONAL ATOMIC ENERGY AGENCY
VIENNA, 2004

IAEA Library Cataloguing in Publication Data

Transition from operation to decommissioning of nuclear installations. — Vienna : International Atomic Energy Agency, 2004.
p. ; 24 cm. – (Technical reports series, ISSN 0074-1914 ; no. 420)
STI/DOC/010/420
ISBN 92-0-114103-3
Includes bibliographical references.

1. Nuclear facilities — Decommissioning. 2. Radioactive substances — Inventories. 3. Nuclear power plants — Decommissioning. I. International Atomic Energy Agency. II. Series: Technical reports series (International Atomic Energy Agency) ; 420.

IAEAL 04-00358

FOREWORD

The transition period between plant operation and the implementation of a decommissioning strategy is a critical one. During this period a number of modifications — both technical and organizational — are needed to adapt the plant to meet new objectives and requirements. It is essential that detailed planning for decommissioning begin in good time during plant operation and preparatory actions for the implementation of the decommissioning strategy be initiated immediately after permanent shutdown. This ensures a gradual transition and minimizes an uncontrolled loss of resources. In some cases, however, this transition process could be better managed. There is significant scope for improvement worldwide. The purpose of this report is to highlight technical, management and organizational issues arising during the transition period; to provide guidance to minimize delays and undue costs; to optimize personnel and other resources; and to initiate preparatory activities for decommissioning in a planned, timely and cost effective manner.

Published information and practical guidance on technical, management and organizational aspects of the transition period are scarce in comparison to what is available on operation or decommissioning. With the growing number of nuclear facilities reaching the decommissioning stage, it is timely to gather and consolidate the experience available globally in a dedicated report. The targeted readership includes all parties involved in operation and decommissioning of nuclear installations, e.g. decision makers, plant operating organizations, decommissioning contractors and regulatory bodies.

A Technical Committee Meeting on the subject was held in Vienna from 4 to 8 March 2002. The meeting was attended by seventeen experts from eleven Member States. The participants discussed and revised a preliminary report written by consultants from Germany, Italy, the Netherlands, the United Kingdom, the United States of America and the IAEA. After the meeting the text was revised by the IAEA Secretariat with the assistance of consultants from the United Kingdom and Germany. The IAEA officer responsible for this publication was M. Laraia of the Division of Nuclear Fuel Cycle and Waste Technology.

CONTENTS

1. INTRODUCTION

The transition period from installation operation to implementation of a decommissioning strategy is an important one. During this period a number of plans and modifications are made to adapt a facility to new objectives and requirements. Transition activities take place between operation and placement of the facility in a safe and stable condition preparatory to safe enclosure and/or dismantling. Typically these activities include defuelling of reactors, retirement of equipment and systems, radiological and waste characterization, operational waste treatment and removal of minor components. Generally, removal or dismantling of major components and, where applicable, safe enclosure (SE) are excluded. However, activities carried out during the transition period will depend upon the type of facility and the regulatory regime. The objective of the transition period is to plan and implement these activities in a timely manner. A cultural change is also needed to reflect different management and working practices. It is essential that planning for the transition and decommissioning begin during operation and that activities be implemented as soon as possible after permanent shutdown to ensure a controlled transition and the best use of resources.

A key to the success of the transition period is the training and preparation of facility personnel. This includes, in particular, utilizing operating staff whose knowledge of the facility and its systems is invaluable during this transition period. In addition, as shown in this publication, a number of strategic and administrative issues need to be addressed before or immediately after permanent shutdown of the plant to support planning for decommissioning and to reduce the burden of operational requirements. Figure 1 provides a possible scheme for decommissioning related activities, projects and organizational aspects covering the period from operation to final dismantling of a nuclear installation.

The subject of the transition period has been dealt with in part by previous IAEA publications, e.g. Refs [1–4], but never addressed as a distinct subject. Little publicly available literature exists on this subject.

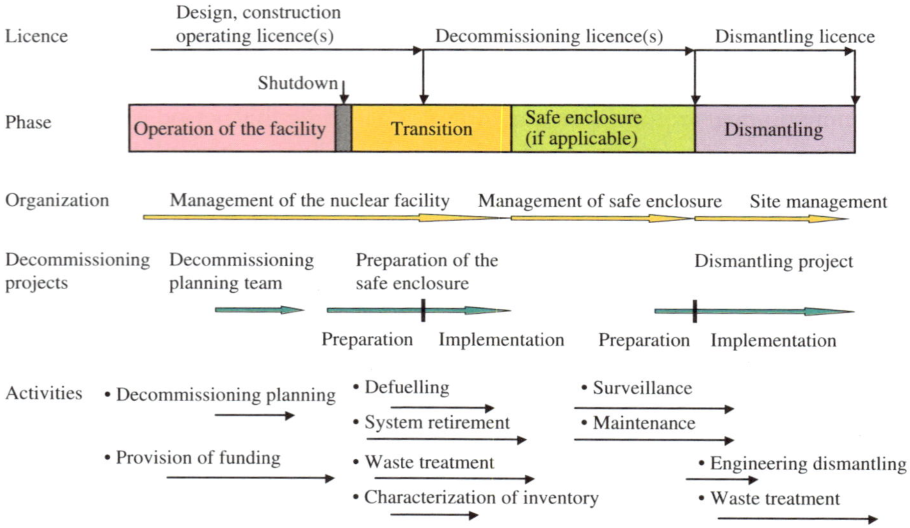

FIG. 1. Decommissioning related activities during the life cycle of an NPP.

2. OBJECTIVE AND SCOPE

The objective of this report is to provide practical advice and information on important aspects of the operation to decommissioning transition with a view to minimizing decommissioning delays and avoiding unnecessary costs during final planning for decommissioning. The readership includes all parties involved in operation and decommissioning of nuclear facilities, e.g. decision makers, plant operating organizations, decommissioning contractors and regulatory bodies. It is important that those responsible for plant life management take account of the issues associated with the transition from operation to implementation of a decommissioning strategy when considering and planning for permanent shutdown (e.g. adjusting the requirements for the retention of spare parts).

This report provides guidance for planned shutdown and excludes post-incident/post-accident shutdown scenarios and abandoned or 'historic' facilities. Other cases of early termination (e.g. as a result of political decisions) are not the primary focus of this publication. However, the activities and processes described may be of assistance in such circumstances. Additionally, early transition planning could provide a basis for any recovery from such a situation. For these facilities, transition may require a significant effort to identify and evaluate existing conditions before identifying appropriate activities and taking the requisite actions. In this case, the transition from operation to decommissioning will assume aspects quite different from a planned shutdown. A complementary publication dealing with management of early termination of NPP operation has been published by the IAEA [5].

This report covers factors applicable to any nuclear installation, e.g. power plants, research reactors, fuel cycle facilities. The guidance provided in this report is generic in nature. Application will necessitate tailoring it to suit the size and nature of a specific facility. For example, a small research reactor will not require the same degree of planning or number of transitional activities as a commercial NPP. Factors specific to various types of nuclear installation are discussed. The focus is on preparatory activities for implementation of a decommissioning strategy. In particular, this report aims to ensure that the transition between operation and implementation of the decommissioning strategy is managed safely, effectively and efficiently. It addresses strategic issues such as planning, administration and implementation. Safety aspects associated with the transition period such as changes to regulations, safety systems or accident analysis, as well as licensing of the transition period, are dealt with in another publication being prepared by the IAEA [6].

3. STRUCTURE

The main part of this publication describes key aspects of the transition period. It emphasizes three topics:

(1) The overall approach and organizational and structural issues (Sections 4–7),
(2) Practical issues relating to planning, management and administration (Sections 8–10),
(3) Technical issues relating to implementation of the transition, including the costs (Sections 11–12).

Section 4 highlights key planning issues and their benefits during early preparation for decommissioning. Section 5 introduces the main goals in the transition from operation to implementation of the decommissioning strategy and gives examples of typical activities. Section 6 highlights issues which could hinder timely decommissioning, i.e. unduly lengthen the transition period leading to an increase in cost. Section 7 describes major changes during the transition period that have to be controlled by plant management. Section 8 describes the organizational and personnel changes during the transition period in more detail. Section 9 turns to strategic issues which have to be managed during the transition period (planning, retirement of systems, technology and tool development, waste management, database development), while Section 10 deals with management/administrative issues arising in parallel (inventory of radioactive and hazardous materials, purchasing policy and retention of spare parts, record keeping, training, and interaction with all stakeholders in the transition and decommissioning process). Section 11 deals with the practical implementation of the issues described in the two previous sections, i.e. spent fuel removal, system cleanout and removal of items, implementation of waste management, decontamination, management of systems, preparation of the facility and buildings for decommissioning. Section 12 gives an overview of cost items which are relevant to the transition period. As a supplement to Section 12, the appendix highlights standardized cost items for decommissioning projects which are relevant to the transition period.

Annex I describes the approaches to and the experience of the transition period in various countries. Annex II provides case studies of problems encountered, solutions and lessons learned during the transition period.

4. PLANNING FOR DECOMMISSIONING DURING THE LIFETIME OF A PLANT

Significant savings can be realized by initiating decommissioning planning in a systematic fashion prior to permanent shutdown and well before a decision to shut down is even made. References [7–9] recommend that a decommissioning plan be produced at the time of the plant's design and construction (see Section 9.1 for more details). The planning should continue while the plant is still operational and information and knowledge of the plant are readily available. This should minimize the need to subsequently track down workers and reconstruct previous events from their memories. It is, however, recognized that for many older facilities such plans will not have been produced at these early stages of their life cycles. A comprehensive, well formulated planning programme would identify the scope of the decommissioning effort required and begin preparation of planning documents. It would also identify and resolve waste management issues; make sound cost estimates to allocate funds for decommissioning; and address the safety aspects, cost and schedule of the decommissioning.

Early planning [10] will ensure that:

(a) Funds have been allocated during the plant's lifetime.
(b) Allocated funds can be spent promptly as needed for a smooth, timely transition to decommissioning.
(c) Time and money will be saved (e.g. due to shortened decommissioning schedules).
(d) Planning is carried out systematically, with less schedule pressure.
(e) Necessary information is available while the plant is operational, records are intact and their location known. Timely access to reliable information can speed up decommissioning planning, reduce uncertainty and risks in the planned work, and result in cost and schedule efficiencies.
(f) Personnel resources (history and expertise) are still available while the plant is operating. However, soon after the announcement of a shutdown, some people will wish to leave. Those leaving might include older, more knowledgeable personnel and younger personnel who would provide a core workforce for later transition and decommissioning tasks. Members of younger and older age groups may leave for different reasons:
    (1) The older group may not wish to leave if they cannot see alternative employment prospects. However, if early retirement packages are offered and accepted by a large proportion of this group, valuable experience and knowledge may be lost. The effects of such losses may be mitigated by implementation of arrangements such as those described in Ref. [11].
    (2) The younger group may wish to leave to seek new opportunities elsewhere. If they do, some of the effort necessary to plan and implement the decommissioning may be lost. To avoid this, steps should be taken to retain some of these people. This could be done by maximizing the opportunities for them to gain appropriate experience and by highlighting the opportunities for future work on decommissioning.
(g) Secure employment for a longer period of time could help overcome the problem of morale where, after plant shutdown, employees are effectively working themselves out of their jobs.
(h) Problem areas such as waste generation, characterization and management will be identified early and plans produced for minimizing delays after shutdown.
(i) While the plant is still operational there is time to plan for decommissioning to achieve the best results, while not adversely affecting operations.

(j) The length of time between shutdown and the start of the decommissioning activities will be considerably reduced.

Early planning can also ease the impact of an unplanned, permanent shutdown. Typically, some older nuclear units have experienced serious technical problems requiring either expensive repairs, upgrades and replacements, or permanent shutdown. Unplanned, permanent shutdowns can have severe economic consequences for the operating company and further complicate the decommissioning effort as priorities have to be changed. Orderly, systematic decommissioning planning is recommended during operation to lessen these effects.

After the plant is shut down, decommissioning may quickly evolve into a ‘material out the door’ mentality to show progress and, as a result, the plan may suffer a schedule squeeze. This may be mitigated by maximizing the number of waste shipments off-site during the operating period, depending on the availability of suitable repositories. If the plant is shut down prematurely, having to undertake decommissioning planning after shutdown may lead to a longer and more expensive decommissioning process.

It should be recognized that assigning the task of planning for decommissioning to an operating organization may cause a conflict with its prime focus of operating safely and economically, particularly where the plant to be decommissioned shares a site with operating plants. To overcome these difficulties, in some countries separate organizational units are in charge of planning for decommissioning. Table I lists possible consequences to the timescale, costs and environmental impact of decommissioning if relevant documentation and/or information is not available.

5. OBJECTIVES OF THE TRANSITION PERIOD

Planning for the transition should begin during the facility’s operating period. Depending on national regulations, an operating licence may remain in effect during all or part of the transition period. The goals during the transition period are to put the facility in a clearly stable condition (e.g. SE), to eliminate or mitigate hazards, and to transfer programme and financial responsibilities from the operating to the decommissioning organization as appropriate. Timely completion of the transitional activities can take advantage of a facility’s operational capabilities before they are lost. Certain activities can be

TABLE I. CONSEQUENCES OF NOT HAVING RELEVANT DOCUMENTATION/INFORMATION FOR THE TRANSITION PERIOD [11]

Data	Consequences
Design, construction, and modifications data	
Site and facility characterization, geological and background radiological data	More time, resources and equipment required; Future litigation due to inadequate data; Significant interaction with the regulatory body on the potential environmental, health and safety issues.
Complete as built drawings, a technical description of the facility, including design calculations	Time/money spent on reconstructing the record and calculations; Direct effect on decommissioning strategy — impact on scheduling; Cannot move to decommissioning without this data being available or reconstructed.
Procurement record and information on the composition of materials used during construction and through the lifetime of the plant	Adequate theoretical assessment of neutron activation of materials is more difficult — waste cost estimates become difficult, leading to considerably more sampling at the facility; Can affect the decommissioning strategy regarding waste management; Causes difficulty in estimating potential dose uptake. This will lead to conservative decommissioning strategies which will effect decommissioning work packages.
Operating, shutdown and post-shutdown data	
Data	**Consequences**
Environmental releases (over the lifetime of the facility)	Lack of assurance on off-site and on-site contamination; Potential for cleanup operations that are not the facility's responsibility; Potential difficulty in releasing land for other uses.
Abnormal occurrence reports	The need to deal with unknowns, can give rise to unexpected operator risk, cause lack of confidence by the regulatory body, the public or the workforce in the management of the decommissioning; Unexpected waste arisings and workforce dose/chemical exposure; Time, costs, resources — can impact the ability to release land.
Records of termination of pipes/cables/vessels	Unexpected hazards arise; Interfere with the development of work programmes — contingency required; Additional wastes generated, e.g. vessels of liquids, cells of material.

Note: All these issues can affect contract bids. Inadequate contingency planning could lead to increased safety hazards, worker dose implications and financial shortfall.

completed more efficiently and cost effectively at this time than if they are postponed.

Activities during a facility's transition period continue to incorporate integrated safety management at all levels to provide protection for workers, the public, the environment as well as the plant. An important objective through transition and eventually decommissioning is to maintain an integrated and smooth process with links to the previous operational phases through reduced surveillance and maintenance.

The decommissioning of a nuclear facility can be supported by the completion of selected activities during the transition period. The transition from the operating to the decommissioning organization can take a considerable amount of time, even years. It is important that progress made during the transition supports the decommissioning strategy. Transition planning is a necessary part of overall decommissioning planning and management.

Key objectives during facility transition are to [12]:

(a) Develop the transition and decommissioning plans, including the specification of end points[1] establishing and defining the required conditions;
(b) Make an expeditious start to activities aimed at eliminating or mitigating hazards, beginning with those that clearly should be carried out regardless of the subsequent decommissioning strategy;
(c) Complete the necessary activities to meet the transition end points, with priority being given to the specified end points for mitigation and removal of hazards and materials;
(d) Maximize the utilization and effectiveness of current operating knowledge, personnel and operating systems or programmes to reduce hazards at the facility, with emphasis on processes and systems for which the skills and knowledge required are unique;
(e) Establish effective relationships among all involved parties, in particular among the operating and decommissioning organization, contractors and authorities;
(f) Mitigate the social impacts of organizational changes;
(g) Reduce the cost of surveillance and maintenance and other transition activities;
(h) Identify the treatment, storage, transport and disposal requirements for all materials and wastes;
(i) Review the budget and funding for specific decommissioning projects;

---

[1] 'End points' are the detailed specifications for the physical condition and configuration to be achieved at the end of a specific phase in the facility's life cycle.

(j) Initiate the ongoing process of culture change and implement new work methods and philosophies.

The degree to which these objectives can be achieved at a facility will vary greatly depending on its function, current condition, configuration and status. High priority is to be given to actions to eliminate or mitigate hazards such as flushing of process systems, removal of waste and defuelling. For other activities, a transition end point development process will ensure that the appropriate activities are identified and completed. Typical transition activities are [12]:

(1) Sale, further use, recycling or dismantling of usable fissile/fertile materials.
(2) Removal of spent fuel and other fissile/fertile material from the plant.
(3) Removal of spent fuel and other fissile/fertile material from the site (if applicable).
(4) Stabilization, treatment and/or removal of potentially unstable materials or wastes.
(5) Reduction or elimination of the potential for fire or explosions from violent chemical reactions or nuclear criticality.
(6) Completion of cleanout operations of systems, lines and other equipment not needed in the future that have the potential for significant radioactive and chemical material inventory.
(7) Neutralization and disposal of hazardous chemicals and oil in storage.
(8) Review, using the safety assessment, of changes in the configuration and status of systems and structures as a result of transition activities, e.g. reducing redundancies in systems and structures.
(9) Revision of operating requirements and controls as appropriate to changed conditions; this should also include the number of personnel required to maintain the appropriate safety standards.
(10) Installation and/or verification of sufficient barriers to prevent the spread of contamination.
(11) Verification of appropriate safeguards and security.
(12) Checking and updating of relevant facility drawings and other documents to reflect changes that have been made during the operational period and/or the transition period.
(13) Training and awareness of facility staff for their future work and roles.

Further detail and comment on end point development and specification are given in Annex I–14, along with references for further guidance. A primary objective during the transition from facility operations through to SE or immediate dismantling is to focus on actions that cost effectively support this process. Experience has shown that a number of general tasks are appropriate during the transition period. These address non-radiological hazards, radiation fields, contamination, waste, isolation and containment, monitoring and control, refurbishment and installation, as well as documenting and labelling of components.

Most, if not all of these apply to facilities that are currently operating, as well as to facilities that have ceased operations and are essentially locked and/or abandoned. The challenge is to identify those transition actions that are appropriate to a particular nuclear facility prior to implementing the decommissioning strategy.

6. ISSUES HINDERING TIMELY DECOMMISSIONING

Experience has shown that the start of a number of past decommissioning projects suffered undue delays and other hindrances resulting in insufficient progress and extra costs. Factors contributing to such delays include:

(a) The unavailability of funds when needed;
(b) Sudden, unplanned, permanent shutdown of a plant (e.g. for political, regulatory or economic reasons);
(c) Lack of a decommissioning strategy or the inability to decide on one, resulting in a 'no action' situation;
(d) Lack of infrastructure (such as waste storage facilities or disposal sites) or developed techniques;
(e) Lack of regulations covering decommissioning;
(f) Loss or demotivation of key personnel and an inability of personnel to adapt to cultural changes;
(g) Little or no planning for decommissioning during plant operation.

The following examples highlight a few critical areas in detail.

(1) A typical issue in the decommissioning of nuclear facilities is the insufficient or non-existent provision of decommissioning funds during plant operation. Except for small, low hazard facilities, e.g. small accelerators or medical laboratories that can be readily dismantled using routine means, a lack of funds severely impacts timeliness, cost effectiveness and ultimately the safety of decommissioning. If nuclear facilities are owned

by the State or State bodies, ad hoc funds are sought from the State budget, often conflicting with priorities in other national sectors. In addition, allocating decommissioning funds in this way may be subject to parliamentary scrutiny and media debate, leading to a perception of low priority and ultimately to undue delays.

(2) Another issue can be uncertainty in the timing and the reasons for permanent shutdown and decommissioning. A long period with no firm decisions on a permanent shutdown and decommissioning strategy could frustrate plant staff and may result in the loss of qualified staff and collective memory. Difficulties in making prompt, clear decisions are often due to intensive lobbying against permanent shutdown for reasons such as expected loss of salaries, fear of staff relocation, or cessation of research, radioisotope production and other programmes. A scarcity of funds to operate the facility or the lack of productive goals (e.g. for research reactors) can also lead to uncertainty.

(3) A related issue is 'taking no action' following the decision to permanently shut down a nuclear facility. Unfortunately, this is common practice, especially for many small facilities that can, by their nature, safely remain in a shut down condition for extended periods. No action is often the result of an incorrect perception that the risks associated with the shut down facility are trivial and can be disregarded. Eventually, this may end with plant abandonment. A policy of doing nothing is generally not acceptable to regulatory bodies and is not recommended by the IAEA.

(4) A fourth relevant issue is the lack of development or availability of decommissioning techniques and waste and material management technologies. In several countries, decommissioning tends to be a 'first of a kind' project with little or no planning or availability of resources. In some countries the importation of techniques and other resources has proved beneficial, but this requires that recipients know how to incorporate these into their decommissioning strategy.

(5) A fifth issue related to a lack of resources and infrastructures is that of decommissioning regulations. In some countries, these are either non-existent or are derived from regulations originally developed for the construction or operation of nuclear installations. Inadequate regulations often result in a convoluted approach, unclear responsibilities and ultimately in undue delays, e.g. regulations for the clearance of sites or materials may not exist or may be inappropriate.

(6) A sixth issue is often the uncertain allocation of roles and responsibilities, as it is well known that decommissioning requires a cultural change. For example, research staff may have difficulties in adjusting to an industrial demolition project. The transition to decommissioning inevitably requires

organizational changes, new lines of reporting and communication, and often the use of contractors. The operating staff who are familiar with routine day-to-day management must now work with and/or manage projects using substantial outside resources. A related problem is the lack of qualified staff due to both the loss of facility staff and a general decline in the nuclear sector.

Issues hindering the timely start of decommissioning were more common at a time when most projects were unique and have become less common now as increasing experience has been gained. An unduly delayed start results in unnecessary expenses, e.g. due to the requirements of the operating licence to maintain full staffing, and equipment maintenance requirements which might otherwise be downgraded in a decommissioning scenario.

Early planning is needed to further minimize the effects of these issues. This implies that high level decommissioning plans are available well before the plant's permanent shutdown. This then leaves sufficient time for the operating organization to optimize the detailed decommissioning plans.

7. MAJOR CHANGES DURING THE TRANSITION PERIOD

The period between the announcement of the shutdown of an NPP and the start of decommissioning can present significant challenges to plant management. They need to prepare for new technical and organizational problems in a climate where there could be pressure to reduce costs and, specifically, the number of staff.

The move towards decommissioning can be regarded as a process of major organizational change which will mostly take place during the transition period. In some projects, attention has largely focused on the technical aspects of decommissioning, with relatively little attention being given to organizational and other personnel issues, in particular an associated significant reduction in numbers of staff. These changes need to be carried out in accordance with rigorous and comprehensive change management arrangements. Table II includes major 'cultural' changes in moving from operation to decommissioning.

Increased levels of uncertainty can threaten staff morale and commitment, and the decision to shut down may itself be preceded by periods of rumour and uncertainty. In an industry where job security has often been

TABLE II. COMPARISON OF THE OPERATING AND DECOMMISSIONING REGIMES [4]

Operations	Decommissioning
Reliance on permanent structures for the operating life of the facility	Introduction of temporary structures to assist dismantling
Safety management systems based on an operating nuclear facility	Safety management systems based on decommissioning tasks
Production oriented management objectives (except perhaps in research facilities)	Project completion oriented management objectives
Routine training and refresher training	Retraining of staff for new activities and skills or use of specialized contractors
Permanent employment with routine objectives	Visible end of employment — refocus of the staff's work objectives
Established and developed operating regulations	Change of regulatory focus
Predominant nuclear and radiological risk	Reduction of nuclear risk, changed nature of radiological risk, significantly increased industrial risk
Focus on functioning of systems	Focus on management of material and radioactivity inventory (e.g. for waste minimization)
Repetitive activities	One-off activities
Working environment well known	Working environment unknowns possible
Routine lines of communication	New lines of communication
Low radiation/contamination levels relatively unimportant	Low radiation/contamination levels important for material clearance
Access to high radiation/contamination areas unlikely or for a short time	Access to high radiation/contamination areas for extended periods
Routine amounts of material shipped off-site	Larger amounts of materials shipped off-site
Relatively stable isotopic composition	Isotopic composition changing with time

taken for granted, this can be unsettling for plant personnel. The plant management may also need to put an early plan in place to deal with the potential social and economic impacts of a plant shutdown. The psychological distress experienced by the workers during the decommissioning of a NPP is described in Refs [13, 14]. References [15, 16] detail the social, economic and environmental considerations in moving from operation to decommissioning at the Trawsfynydd nuclear power plant in the United Kingdom. These references also describe a public consultation process and the efforts made to ensure that the local population was well informed and understood the issues involved.

During the transition, plant management may use contractors to make up for any shortfalls resulting from the loss of experienced staff, and also to bring in the specialized skills required throughout this period. However, it is vital that the licensee retain enough suitably qualified and experienced personnel to understand and work toward the plant's safety, and to be an 'intelligent customer' of these contractors. This is especially important during the transition period if the numbers of permanent staff are declining.

Older plants may not have a comprehensive set of drawings and procedures. Many historical aspects of plant design and operation which need to be accessed during the transition period are known only to individuals and are not recorded in documents. These people are therefore important during the transition period when their knowledge and experience may be required. This experience should ideally be documented in a form that is available for use by other personnel. Reference [17] is a comprehensive study of the challenges to culture, morale and skills during the transition from operation to decommissioning.

8. ORGANIZATION AND PERSONNEL

The decommissioning of a large nuclear facility with the activities involved in the transition is a major project. The best project management practices, tools and techniques, as well as quality assurance processes, are vital. The organizational aspects of the preparation for implementation of the decommissioning strategy are dealt with in the following sections. Further guidance can be found in Ref. [4].

8.1. PREPARATION FOR THE TRANSITION PERIOD

It is important to establish a project team to plan for the transition and decommissioning well in advance of the final planned shutdown. This team does not need to be large or employed full time and could also be a unit separate from the operating organization. Its technical and safety expertise should include knowledge of system reconfiguration or retirement, spent fuel and waste management, plant history, licensing and other decommissioning aspects. Standard project expertise such as cost estimation, time and work scheduling are also important.

This team may report to senior management but not be responsible for the day-to-day operations of the plant. It will be responsible for updating the decommissioning plan based on the agreed decommissioning strategy. Typical objectives of the project team could include:

(a) Project development (time, cost and quality),
(b) Cost estimate verifications,
(c) Project risk evaluation,
(d) Plant system reconfiguration and retirement,
(e) Spent fuel management options,
(f) Waste management plans,
(g) Preparation of safety documentation,
(h) Interaction with stakeholders,
(i) Staffing plans,
(j) Specification of transition end points,
(k) Management of records,
(l) Implementation of the change management strategy.

Establishment of accurate decommissioning costs and risks provides important information for allocating and managing the decommissioning fund. The cost of a small project team could be considered as an investment to achieve a better managed decommissioning project.

A senior transition/decommissioning project manager who has the required skills, qualifications, experience and delegated authority should be appointed before a planned shutdown. This manager, in consultation with the management of the operating organization, sets up the decommissioning project management team which will develop details of the transition and decommissioning. This team will expand and contract as required during the project.

8.2. ORGANIZATION AND MANAGEMENT DURING THE TRANSITION PERIOD

During the transition period many operational hazards are removed in preparation for SE and/or immediate dismantling. This may include removal of the spent fuel, draining of systems, post-operational cleanout (POCO)[2] and removal of waste generated during operation (see Section 11). The management structure will at all times reflect the circumstances and continuing responsibility of the licensee for the licensed site.

At the start of the transition period, the organization will inevitably be that which ended the operational phase. Even in cases where a new operating organization takes over for decommissioning, it is likely that most of the operating staff will be retained and their roles will change to reflect the activities during the transition period, as depicted in Fig. 1. It is essential that the organizational changes and plant modifications be well defined prior to shutdown. It is important that these changes address roles, responsibilities and reporting lines. Figure 2 shows a typical functional[3] organization as it might be

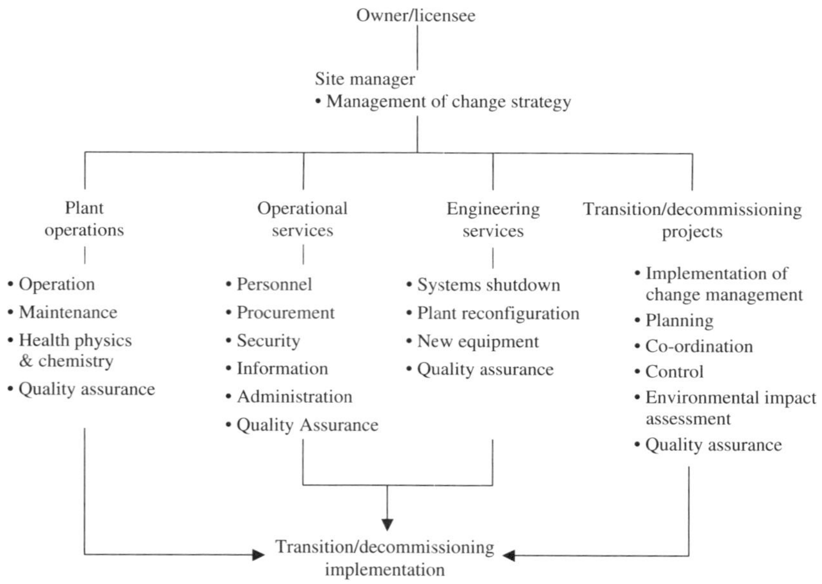

FIG. 2. Example of a functional organization during the transition period.

---

[2] This terminology is in general use in the UK.

[3] "Functional" means that it indicates types of activities, but not necessarily lines of reporting.

modified for transition and decommissioning projects and tasks. In addition to the facility personnel, contractors will be assigned to some jobs, particularly during dismantling. Figure 3 shows the staff reorganization during the transition period at the Trino NPP in Italy.

8.3. THE IMPACT OF THE TRANSITION AND DECOMMISSIONING ON HUMAN RESOURCES

There will inevitably be constraints on the approach to staffing during decommissioning. In some facilities staff numbers are likely to be held close to operating levels (e.g. in reactor plants until the fuel has been removed). In other plants the change will depend on the need to stabilize or remove existing hazards. The number of operating personnel needed will eventually fall as shown in Fig. 4. Figure 4 (a) illustrates the general trend in staffing levels as the facility is shut down, during the post-operation transition, when the facility is placed in a condition of SE, and during final dismantling. The reduction in operating staff as systems are retired and licensing conditions are reduced is somewhat compensated for by additional staff being required for preparation of the SE and final dismantling. Figure 4 (b) illustrates the trend for a facility proceeding to dismantling soon after shutdown. In this situation, a significant increase in staffing levels is needed for the decommissioning activities. A number of basic points need to be addressed and decisions made concerning the following [4]:

(a) The required organization,
(b) A staff reduction profile,
(c) The use of operating staff to undertake decommissioning project tasks,
(d) Sharing of key resources among plants,
(e) Policies for choosing which work will be contracted.

The staff reduction profile will depend on the numbers, qualifications and experience of the personnel needed for the actual work to be carried out. Having established such a profile, commitments can then be made to staff regarding the length of their remaining employment and the implementation of staff reduction measures can be monitored. During the reorganization at Trino (see Section 8.2), the main changes were:

(1) The number of personnel in the operations department was reduced to the minimum level on shift required by law when irradiated fuel is still in the pool;

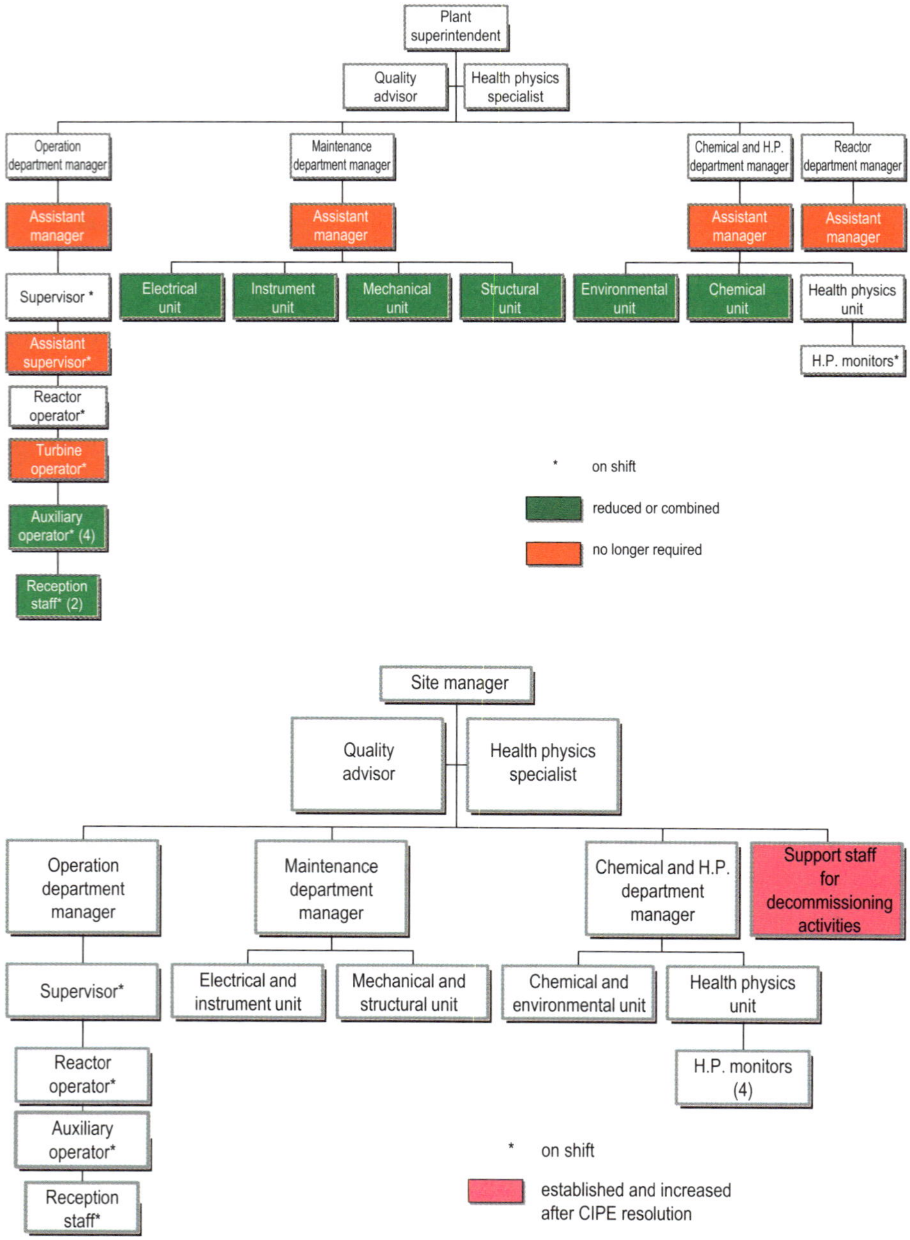

FIG. 3. Organization of the Trino NPP (a) before and (b) after restructuring during the transition period.

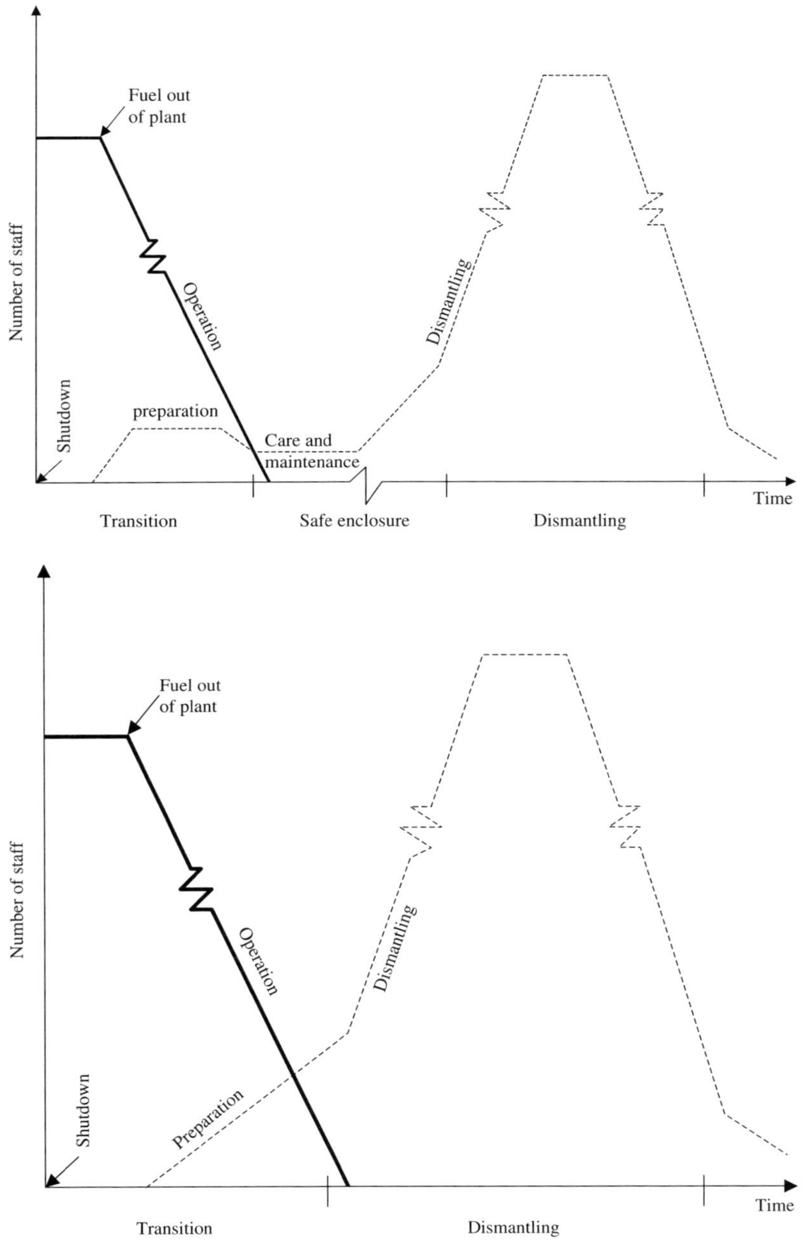

FIG. 4. (a) Staffing trend during the transition, SE and dismantling; (b) during the transition to immediate dismantling (not to scale).

(2) The maintenance department and the chemistry and health physics department were reorganized to create combined sections which contributed to the production of decommissioning documents;
(3) The health physics personnel were taken off shift and focused, during normal hours, on pre-decommissioning activities such as radiological characterization and waste treatment;
(4) The professional staff in the technical area were focused on preparation of the documents needed for decommissioning.

Retaining a large number of operating staff will inevitably mean that they undertake decommissioning tasks. They will require retraining in new skills and reorientation of their attitudes towards decommissioning, e.g. system isolation, dismantling, draining and flushing, waste characterization, dismantling and size reduction techniques, etc.

Early planning with regard to the timing of final shutdown and the selected decommissioning strategy plays a major role in facilitating the management of personnel relocation and the retention of key staff. However, the use of outside contractors to perform the majority of the decommissioning activities may lead to resentment and demotivation among the existing plant staff.

In order to focus on the completion of tasks it is helpful if arrangements are in place to warrant subsequent relocation of staff to other plants, projects or similar organizations, or out-placement to other job markets. One way of approaching this would be to form teams of skilled, experienced personnel who could provide services to other similar plants as effectively as contractors [4].

It is important to provide appropriate incentives to the remaining staff (and contractors) to work effectively and in a manner that safely maintains the decommissioning programme's schedule, quality and budget. These incentives may differ from situation to situation and, while seeking to encourage a safe adherence to the decommissioning programme, should encourage staff to strive toward completion of the work and mitigate concerns about future employment.

9. STRATEGIC ISSUES

This section describes selected strategic issues for which pre-shutdown planning should be conducted so as to arrive at decisions on courses of action upon shutdown. This includes overall planning for the transition and implementation of the decommissioning strategy, cost reduction, waste management issues, and development of techniques and tools required for decontamination and dismantling. An example of the activities required for a specific transition to decommissioning project is given in Ref. [18], which includes licensing, engineering, implementation, staffing levels and overall organization.

9.1. OVERALL PLANNING FOR THE TRANSITION PERIOD

Section 8 recommends establishing a team that:

(a) Updates the decommissioning documentation, including the decommissioning plan as required by IAEA Safety Guides [7–9];
(b) Develops plans for the transition from operation through shutdown and decommissioning. This is entirely consistent with the following paragraphs extracted from Ref. [7], which calls for the decommissioning plan to be a 'living document' throughout the plant's life cycle:

> "5.6. An initial plan for decommissioning should be prepared and submitted by the operating organization in support of the licence application for the construction of a new reactor. A generic study showing the feasibility of decommissioning may suffice for this plan, particularly in standardized installations. Depending on applicable regulations, the plan should address the costs and the means of financing the decommissioning work.
>
> 5.8. During the operation of a reactor, the decommissioning plan should be reviewed, updated and made more comprehensive with respect to technological developments in decommissioning, incidents that may have occurred, including abnormal events, amendments in regulations and government policy, and, where applicable, cost estimates and financial provisions. The decommissioning plan should evolve with respect to safety considerations, based on operational experience and on information reflecting improved technology. All significant systems and structural changes during plant operation should be reflected in the process of ongoing planning for decommissioning.
>
> 5.9. When the timing of the final shutdown of a nuclear reactor is known, the operating organization should initiate detailed studies and finalize proposals for decommissioning. Following this, the operating organization should submit an application containing the

final decommissioning plan for review and approval by the regulatory body. The decommissioning plan may require amendments or further refinements as decommissioning proceeds, and may require further regulatory approval."

The activities carried out during the transition period have two main objectives:

(1) The efficient operational conversion of the facility from its original mission to one in which operations, surveillance and maintenance are reduced, consistent with the lower safety risk, the systematic reduction in hazard and the need to cost effectively prepare for either SE or immediate dismantling.
(2) The preparation of a detailed decommissioning plan, which requires the most current information available regarding the condition of systems, structures, components and materials. For example, a full radiological characterization is required to provide data for updating the decommissioning plan and the waste management strategy (Sections 10.1 and 11.5).

Such activities are normally initiated well before shutdown, including selection of a decommissioning strategy which might, however, be finalized only after shutdown. Regardless of how far the decommissioning strategy has advanced, transition activities serve to achieve a safer and more economical configuration, e.g. by reducing the inventory of radiological material, dealing with spent fuel, removing hazardous chemicals, focusing surveillance and maintenance on plant features needed to control contamination and other hazards.

9.2. COST REDUCTION BY RETIREMENT OF SYSTEMS OR RECONFIGURATION

During the transition period, activities are planned and carried out which lead to simplified operation, reduced surveillance and maintenance requirements and lower operating costs. This can be achieved by identifying those plant systems which will become redundant after final shutdown. Further consideration should be given to systems that are needed after shutdown but which are costly to operate and maintain, e.g. the capacity of the ventilation system needed to control contamination in shut down facilities can be greatly reduced. Cost reductions will also take place as a result of changes to technical specifications as the licence is amended. Implementation aspects are addressed in detail in Section 11.6.

Cost savings can be achieved from reductions in:

(a) Labour,
(b) Power and fuel consumption,
(c) Consumables,
(d) Surveillance and maintenance,
(e) Regulatory and technical requirements including inspections,
(f) Training,
(g) Recycling of material and components,
(h) Nuclear insurance.

Figure 5 shows the trend in power consumption for the Trino NPP in Italy. Once NPPs are shut down, nuclear materials continue to be kept under strict supervision and only the movement of fuel elements off-site is foreseen. Security costs and nuclear liability insurance can be reduced when NPPs are shut down and fuel elements are moved off-site.

Similarly, systematic hazard reduction in other types of nuclear facility may lead to a reduction in liability [19]. However, insurance liabilities for

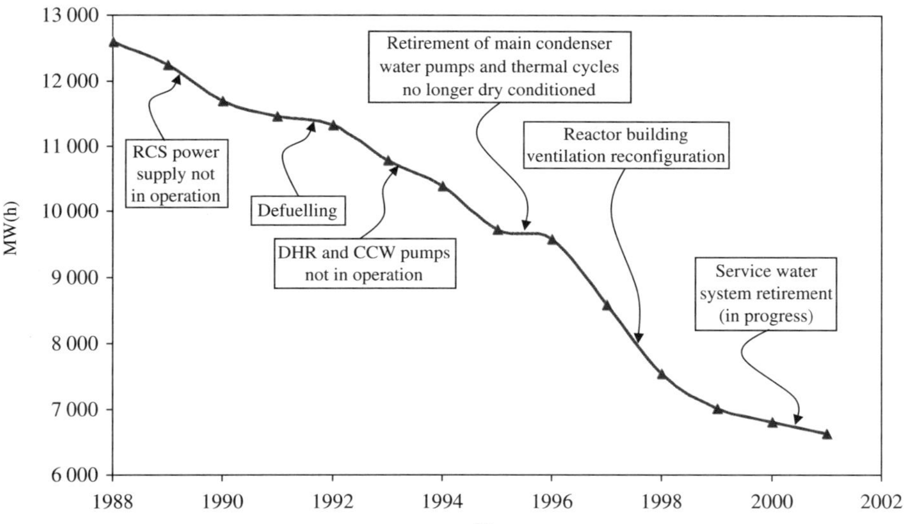

FIG. 5. The electricity consumption trend for the Trino NPP after shutdown (DHR: decay heat removal; CCW: component cooling water; RCS: reactor control system).

conventional industrial risks may increase due to extensive dismantling of the inactive plant and a greater potential for environmental contamination from hazardous materials, e.g. asbestos, oil, PCBs.

9.3. DEVELOPMENT OF TECHNIQUES AND TOOLS

Decontamination and dismantling techniques and tools for nuclear facilities are widely available, as has been demonstrated by the successful completion of a large number of decommissioning projects. Most decommissioning tasks, especially concerning reactors, can be accomplished using existing techniques for which expertise is already available. However, there may be tasks that will require the development of new techniques or the adaptation of existing techniques, in particular for the decommissioning of facilities with special design features, experimental and prototype facilities or those with an unusual operating history (e.g. after incidents, extensive reconstruction, etc.).

While planning for decommissioning and, in particular, the transition period, it is important to identify whether all the planned tasks can be completed using existing techniques and tools, or if the development of new or the adaptation of existing methods and techniques is needed. It is desirable that this development or adaptation be started during the transition period (in laboratory, mock-up, pilot or full scale) in order to have the decontamination and dismantling techniques available when work commences. The development activities will depend on the chosen decommissioning strategy and selection of the best methods for decontamination, size reduction, dismantling, demolition, waste packaging, etc. On the basis of test results and demonstrations, decommissioning plans can be optimized and finalized. One example of extensive full scale mock-up training for dismantling purposes is described in Ref. [20]. Another example (Fig. 6) shows the mock-up test of the manipulator (called DENAR) in preparation for remote decontamination at the A-1 reactor decommissioning project in Bohuniče, Slovakia.

A related topic is sampling, characterization and location of contaminated materials and areas where decontamination is required in support of decommissioning. It may also be necessary to test decontamination techniques on selected areas of the plant and its components. Similarly it may be possible to measure activated samples from a reactor to validate computer estimates with actual measurements in order to optimize size reduction, waste shipping and disposal. The transition period provides the opportunity for any additional sample collection and plant characterization (see Sections 10.1 and 11.5).

9.4. WASTE MANAGEMENT

Activities during the transition period have the potential to increase both the volume and the variety of wastes generated. Planning should ensure that there is sufficient capacity for the treatment of these wastes, their storage or transport and disposal. The issues that may need to be considered include:

(a) The wastes that will arise during the transition period;
(b) Wastes held in interim storage at the facility which need to be recovered for treatment, conditioning and disposal (e.g. sludges, ion exchange resins, spent radiation sources, scrap components);
(c) Long term storage requirements;
(d) Wastes from decontamination and cleanup operations (e.g. additional resins from chemical decontamination, demolition wastes, etc.);
(e) Availability of disposal routes, including transportation;

FIG. 6. Mock-up test of the manipulator at the A-1 reactor decommissioning project in Bohunice, Slovakia.

(f) Materials and equipment left over from experimental and research programmes;
(g) Waste retrieval and conditioning methods;
(h) Waste characterization programmes and techniques;
(i) Waste minimization programmes and techniques;
(j) Clearance levels;
(k) Regulatory authorizations.

As an example, Fig. 7 shows a drum monitoring station for clearance purposes installed for the A-1 reactor decommissioning project in Slovakia. In addition to radioactive wastes, it is possible that significant amounts of hazardous material and unwanted chemicals will also require packaging and disposal. It would benefit the planning and implementation of the transition and decommissioning activities if, during the operational period, as much waste as possible were removed from the facility for disposal. Spent fuel management is dealt with in detail in Section 11.1.

FIG. 7. Drum monitoring at the A-1 reactor, Slovakia.

10. MANAGEMENT/ADMINISTRATIVE ISSUES

This section highlights selected management and administrative issues that need to be considered during the transition period such as an inventory of hazardous material (including the radiological inventory), the purchasing and spares policy (with a focus on cost reduction), record keeping and interaction with all relevant stakeholders in the decommissioning process. These tasks may take on a high priority when the transition period is being planned.

10.1. TAKING A COMPREHENSIVE INVENTORY OF RADIOACTIVE AND HAZARDOUS MATERIALS

The objective of the characterization of radiologically and chemically hazardous materials is to provide a reliable database on the quantity, types, distribution, and physical and chemical states of these materials. This should include contaminated land. Characterization includes reviewing existing data and calculations, taking in situ measurements, sampling, analysis and undertaking of further calculations as needed. This provides a significant input to the decommissioning planning process and the development of successful implementation plans. Information should be updated on a regular basis to account for waste disposal, material removal, radioactive decay, etc. It is crucial that the database remain available during SE and dismantling (see Section 10.3). This information will aid decisions for partial or full decontamination, provision of shielding, partial removal of equipment, waste classification, etc. Further information can be found in Ref. [21].

10.2. REVIEW OF PURCHASING POLICY AND SPARES

Expenditures can be greatly reduced if the purchasing and spare parts policy is carefully reassessed. Many purchasing contracts for components, consumables and services are subject to high quality standards related to the requirements of operating plants in the nuclear industry. Also, retirement of systems leads to a reduction in the need for spare parts.

Purchasing contracts can be re-evaluated for their applicability and justification in the forthcoming decommissioning phase. Many components and consumables do not necessarily have to meet the same quality standards as required during the operational life of the plant. On the other hand components already in stock and meeting these standards could be used in other plants and sold as such. It should be emphasised, however, that to demonstrate that these components meet the standards, the full documentation must be in place and the component must be in demonstrably good condition, e.g. kept under suitable storage conditions. The potential also exists for reuse of refurbished components if required quality standards can be achieved. The policy on stocking of components and consumables should be reviewed as well. Component stock size requirements can be reduced in many cases (e.g. where short delivery times are no longer required) and completely lifted in others.

Unused fuel assemblies in NPPs will also be removed from the plant during the transition period. In a few cases this fuel can be sold to be used in other (equivalent) plants, but in many cases the fuel design will be plant specific and therefore of no use to other power plants. In some countries it is possible to return the fuel to the manufacturer where it can be reprocessed. Similar considerations apply to uncontaminated chemicals, oil, etc. Another example is the selling of spare control rods after dismantling (see Annex I–6).

10.3. RECORDS TO SUPPORT DECOMMISSIONING

Prior to shutdown and during the transition period, collection of information and records that will be needed to support decommissioning plays an important role. Several challenges related to record related decisions are briefly described here.

(a) When to assemble the collection of records and the database: The late creation of a full set of essential decommissioning records may cause difficulties due to reduced availability of time, resources and personnel. As a minimum, it is important that which records will be needed and their location be identified prior to shutdown. This should be supported by a suitable records management system.

(b) Future retrievability: Record keeping in a deferred dismantling scenario poses long term issues with respect to both degradation and retrievability. In this case, reliance will be completely dependent on records assembled several decades earlier. In particular, any records stored in electronic formats and media need to take into account future changes to systems. Paper or film records are subject to ageing. Record storage systems will have to meet national requirements.
   (1) How will they be maintained: Questions that need to be addressed include:
      (i) Which organization will be responsible for keeping records?
      (ii) Will there be both central and local copies?

(iii) Who will have access before they are needed for decommissioning?
(iv) What type of database will be used?
(v) What are the quality assurance requirements?

(2) Which records are to be retained: Selection criteria will depend on future needs. A key factor is whether the decommissioning strategy is to be immediate or deferred dismantling. Immediate dismantling is less problematic because the location of much of the information needed will be known and readily available. Selection criteria are based on:
(i) Technical and safety support (radiological and industrial) for decommissioning activities,
(ii) Technical and safety support for surveillance and maintenance during SE,
(iii) Compliance requirements for statutory and regulatory instruments, including dose and health records,
(iv) Historic or social interest,
(v) Defence against litigation.

(c) Types of record: Application of the above criteria may still require further focus on specific needs, such as inaccessible areas. Certain areas may not have been accessible during normal operations, but workers may need to access these during decommissioning operations. Knowledge of the radiological conditions in these areas, e.g. around the reactor or within the biological shield, will help to minimize occupational exposure during decommissioning. It is also important that the full spectrum of the material characterization and information on the structural condition at shutdown and the end of the transition period are known and recorded. Most of the currently available characterization techniques are suitable for direct electronic recording (Figs 8, 9).

Detailed information on how to develop and maintain a set of records for decommissioning purposes is provided in Ref. [11].

10.4. INTERACTION WITH STAKEHOLDERS

The licensee and the project team can best perform their duties by early interaction with those stakeholders[4] who have an involvement or interest in the

[4] A stakeholder is a person or group who can affect or is affected by an action.

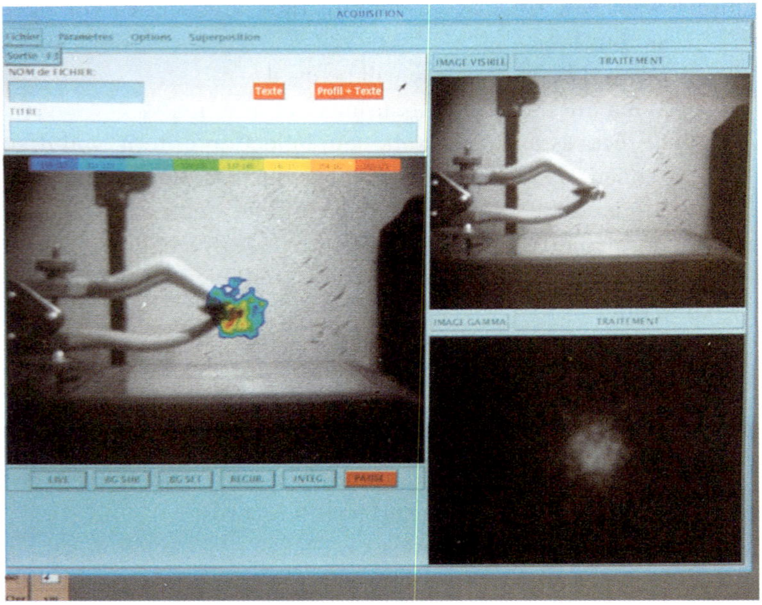

FIG. 8. ALADIN, France: a gamma camera to identify hot spots (red in the photo).

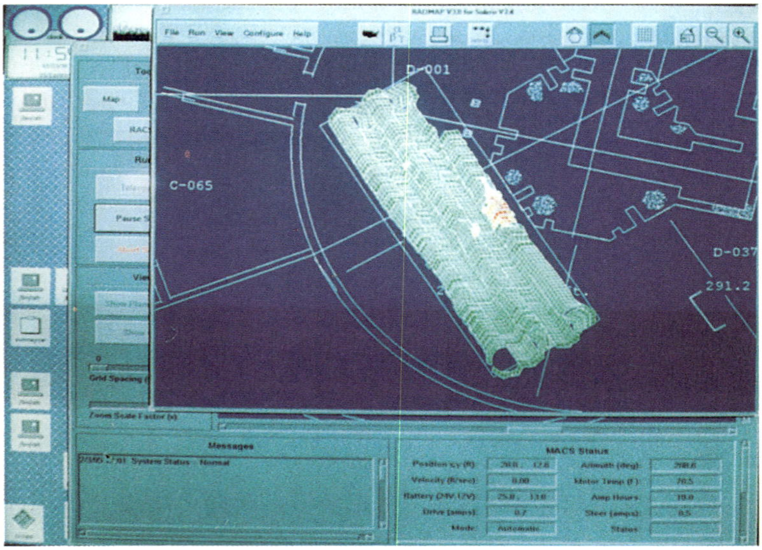

FIG. 9. Mobile automated characterization system (MACS) (CP-5 decommissioning project, Argonne National Laboratory, USA): graphic result of floor survey; yellow and red areas indicate contamination (Argonne National Laboratory).

transition and decommissioning process. Such interactions may occur at all levels within the licensee's organization. Stakeholders typically include:

(a) Regulatory authorities: Nuclear safety, transportation, environmental and radiation protection;
(b) Local, regional and national governments;
(c) The general public: individuals, communities, pressure groups and media;
(d) Employees;
(e) Shareholders;
(f) Labour unions;
(g) Contractors;
(h) Waste management organizations;
(i) The nuclear industry;
(j) National standards groups and professional societies;
(k) International organizations.

The degree of influence of and the priority given to these stakeholders will depend on the individual facilities and local circumstances. Stakeholders provide technical, social, economic, environmental, regulatory and legislative input into the process. Their involvement can provide a valuable review function and constructive input to the project team. It is important that these interactions be initiated as early as possible and developed through the transition period. Examples of activities that may be appropriate include:

(1) Regular meetings with labour unions/employee representatives,
(2) Production of project summary literature for public distribution,
(3) Presentations to the stakeholders,
(4) Public meetings/consultations,
(5) Media interviews and press conferences,
(6) Presentation of papers at local and international conferences.

It is necessary to be aware of issues such as local sensitivities, security arrangements, possible reuse of the site, the potential for official public inquiries and the need to meet all regulatory requirements. The importance attached to different issues varies significantly with the cultural values and background of individual countries and decision making groups. It is also affected by the social and economic status of the country, groups and individuals, educational levels, demographics, individual and group life experience, etc. These can all affect the relative acceptability of various proposals.

In some countries, organizations engaged in decommissioning have endeavoured to reach a consensus among the various interested parties by providing information concerning the contaminated sites, the associated hazards and the proposed methods of dealing with those hazards. This information may be enhanced by general education on radiation and contamination so that the decisions can be more readily understood. This is particularly relevant during the transition period when decommissioning issues come to the attention of the public. Experience suggests that the quality and particularly the acceptability of the subsequent decisions are improved by this process [22].

Public acceptance of major changes can be important for their successful implementation. This process can be assisted by public input while strategic options are being considered, as well as by public consultation, depending on the country's legal system. Strategic options that are likely to lead to public acceptance are sometimes preferred as they also build up public trust and confidence.

There may also be concern about the impact on local employment. If the site is in operation when the closure decision is made there will usually be issues related to changes to the local employment base. These changes could be caused by a changing mission (i.e. from operation to decommissioning) and the potential future release of the site/facility for new purposes. Resolution of this issue is dependent on local customs and conditions. A case where extensive public consultation resulted in modification of the decommissioning strategy is comprehensively discussed in Ref. [16].

10.5. TRAINING TO SUPPORT THE TRANSITION

The extent of training of personnel to support the transition will depend upon the activities undertaken. If no dismantling or new activities (e.g. POCO) are to be undertaken, training will be specific to the changing conditions of the facility and the differences between normal operations and permanent shutdown. However, training may be required for the dismantling of a non-active plant and the introduction of novel techniques for dealing with wastes. Dismantling of non-active plants can be used to train personnel for the future dismantling of active plants. Training material for personnel assigned to a specific facility should be based on the following considerations:

(a) Facilities currently in operation that are going to be held in a shut down condition: much of the knowledge that was required for past outages, maintenance, refuelling, modernization and modification will be needed;

(b) Facilities which are to be shut down in preparation for SE or dismantling may require the development of skills in such areas as preliminary plant cleanout, waste conditioning and dismantling activities;

(c) Facilities that have been out of operation for an extended period and that require inspection to determine whether additional preparatory work is needed prior to decommissioning: training in this situation will require gaining familiarization with the extant conditions.

Some training subject areas for the transition are:

(1) Management: Emphasize project as opposed to production/operational management principles. This will involve training in technical, cost and schedule preparations. Training should ensure that management personnel are familiar with the concepts of:
   (i) Determining the criteria and conditions governing staff reductions,
   (ii) Amending the safety assessment,
   (iii) Cost estimation and budgeting,
   (iv) Complementing the operating organization, e.g. by use of contractors,
   (v) Change management, e.g. arrangements to deal with staff reductions, etc.

(2) Safety analysis: Training should focus on safety issues for a facility that is no longer in operation. It should also address how the safety conditions could be changed and how the requirements for technical specifications, surveillance and maintenance could be reduced.

(3) Plant engineering: Training of personnel responsible for operation of systems and equipment should focus on:
   (i) Shutdown and isolation of systems,
   (ii) Determination of the required level of surveillance and maintenance as a result of changes in the safety case. The latter should emphasize both cessation of activities or reduction in their frequency, recording any system changes and tagging and identifying systems.

(4) Inspection of orphaned facilities: Where facilities have either been abandoned or shut down for an extended period of time, the training of inspectors should highlight structural assessment (building and plant), roof integrity evaluation and identification of radioactive, chemical, electrical, and other physical hazards.

(5) Cost estimation: As reflected in the appendix, many of the cost line items for transition are not normally considered during operation. Those responsible for budgeting should become aware of the differences as well as models used for estimating such costs.

(6) Waste management: New waste characterization, waste retrieval and conditioning techniques may be developed, for which training is required.
(7) Technical and manual work: Training should emphasize implementation of many of the above subjects (for example permanently isolating systems, changes in surveillance and maintenance, new techniques).

11. IMPLEMENTATION ISSUES

This section deals with actual operations that are normally carried out during the transition period. These include spent fuel removal, draining and drying of circuits and systems, preservation of equipment, waste removal, waste management, removal of components and system management (e.g. reduction or modification of ventilation systems). The removal of combustible materials, radioactive materials and hazardous chemicals will reduce the potential source term for any potential accident and reduce the hazards. It is important to ensure that the above activities are carried out using trained personnel, with appropriate approved procedures and all engineered safety features in place.

11.1. REMOVAL OF SPENT FUEL

Experience has shown that the removal of spent fuel is a very important step in the decommissioning of reactors. The preferred solution is the early removal of the spent fuel to a storage facility, to a reprocessing plant or to a disposal facility. Benefits of early defuelling include decreased radiological hazards, timely implementation of dismantling, downgrading of the operating licence, shutdown of some systems (e.g. cooling water, surveillance), and reduced safeguards requirements. In addition, as long as fuel remains in the fuel storage pools, continuous manning of the unit with shift workers may be required, albeit with a reduced number. If consideration is given to adopting shorter refuelling cycles towards the end of the plant's life, the period required for cooling the fuel in the fuel storage pool is reduced. Thus the pool can be emptied earlier than would otherwise be the case, reducing costs.

As long as all infrastructure and provisions are in place, defuelling can be done as during plant operation. However, if removal of the fuel is delayed for a very long time, loss of qualified staff and necessary equipment could become a problem. In some research and prototype reactors, defuelling is not a routine operation and requires special planning during the transition period. For example, depending on the reactor type:

(a) No fuel storage pond may be available,
(b) Lifting equipment may not be capable of carrying fuel transport containers,
(c) Space may not be available for loading fuel elements into transport containers.

Developments worldwide have resulted in a situation where removal of spent fuel to off-site facilities may become a serious problem. For example, many NPPs and research reactors have been provided with fuel by a supplier from another country. Reactor operating organizations may have planned to return the spent fuel to the supplier which, however, in many cases will have become impracticable. As this situation was unforeseen, only a few of these reactor operating organizations have their own off-site spent fuel storage facilities. In other cases, plans for a national fuel disposal facility have been seriously delayed [23, 24].

Therefore it is important for transition planning to consider what is to be done with the spent fuel. It may be necessary to consider constructing a spent fuel storage facility if no other alternative exists. This may have to be considered on a local, national or regional level. Currently, some Member States consider the use of a spent fuel storage installation that is remote from the reactor and which uses dry or wet storage technology (e.g. casks, modules and vaults) to be a successful method of storing spent fuel after sufficient time has elapsed for decay heat reduction (Fig. 10).

In addition to the removal of nuclear fuel it is very desirable to eliminate the possibility of criticality during the transition period. If the spent fuel and other nuclear materials cannot be moved outside the nuclear installation, decommissioning cannot be fully completed.

11.2. SYSTEM CLEANOUT OPERATION

It is essential that process and auxiliary fluids from redundant systems be removed and disposed of while personnel are available who are trained and qualified to operate that equipment. After removal, the systems should be flushed until residual contamination is below predetermined criteria and dried as appropriate. The criteria should be based on (a) regulations, (b) an assessment with respect to future decommissioning worker safety, or (c) limiting degradation (e.g. caused by corrosion) while in SE.

An example (SGHWR, UK) of draining and flushing auxiliary systems connected to the primary circuit followed by treatment of the resulting liquid waste by the installed waste treatment systems is given in Ref. [25].

Experienced personnel are also needed to deal with radioactively or chemically contaminated solids. This is particularly important when the handling equipment is immediately available. Important examples are materials remaining in hot cells that have working manipulators, materials in storage that require such hot cells for handling, items that are in ponds for shielding reasons, and alpha emitting items that require glovebox handling. Again, it is important that the knowledge of current plant workers (with regard to operating the equipment and/or being familiar with the characteristics of the material) be used to the maximum. It is also important that such operations are not unduly postponed even when handling equipment is not immediately available or not working. In such cases, devising alternate removal means during the transition period is a priority.

Organic fluids or hazardous chemicals used during operation, e.g. lubricants, hydraulic oil, acids, etc. are removed and disposed of during the

FIG. 10. CANSTOR^{TM} concrete module for interim dry storage of CANDU type spent fuel, Gentilly-2 NPP, Canada.

transition period. Radioactively contaminated organic and flammable fluids, as well as non-radioactive hazardous fluids (e.g. PCB transformer oil) or solids (e.g. asbestos) will require special disposal procedures.

11.3. TREATMENT, CONDITIONING, STORAGE AND/OR DISPOSAL OF WASTE DURING THE TRANSITION PERIOD

Most wastes generated during the transition period are similar in nature to those produced during plant operation and maintenance [26]. At the end of the operational life of a facility, effort is generally directed at the removal or reduction of any hazard in all areas of the plant to provide a 'passive' safe environment during SE. The amount of work to be undertaken will depend on the operations that were carried out within the facility and the nature of the hazardous inventory associated with the process, that is radiological, toxic or non-hazardous.

Such removal or reduction is important for the transition period although historically this has frequently been delayed until the start of dismantling. However, it should be emphasized that if POCO is deferred until the dismantling phase the associated risks remain and are transferred to the future. Methods for assessing the overall requirements for cleanout, both in terms of the need for and extent of such operations, are given in Ref. [27]. Ultimately, if significant costs or personnel exposure are involved, the decision making process will be based on the overall net benefit.

At final plant shutdown, all waste remaining from past activities is commonly removed from the plant for treatment, conditioning, packaging and storage or disposal. As an example, Fig. 11 shows the installation of a cementation plant for liquid waste in preparation for the decommissioning of the Salaspils Research Reactor, Latvia. Waste management includes not only process fluids (Section 11.2) and sludges but also solid waste (e.g. trash, insulation, loose tools) from controlled areas. The latter can comprise a significant number of items, e.g. in research facilities experimental equipment has often remained in the building years after its use (Fig. 12). Special attention should be paid to long neglected areas.

11.4. DECONTAMINATION OR FIXING OF CONTAMINATION

Decontamination after the end of operation will help to reduce occupational exposure during future decommissioning activities. Decontamination

FIG. 11. Installation of the cementation plant in preparation for decommissioning of the Salaspils IRT reactor, Latvia.

FIG. 12. The reactor block and experimental test apparatus of the IRT reactor, Georgia. At the end of operation, a research reactor's hall is typically full of experimental devices which are removed in preparation for decommissioning.

may be necessary in the circuits, tanks and containers to remove the activity from inner surfaces, as well as on the surfaces of components and buildings to reduce the potential for airborne contamination. Fixing activity on accessible surfaces may be a viable alternative to its removal. However, implications for eventual dismantling and handling of such material require special consideration and specific recording.

11.4.1. Deciding on the need for and the extent of decontamination

In general, decontamination that is carried out during the transition period is primarily aimed at dose reduction and is not intended for material clearance. Aggressive decontamination methods can often be applied where the systems are no longer needed for operation.

The decision whether to decontaminate a nuclear facility (or parts of it) will in general depend on the type of plant, the radionuclide vector/inventory and other constraints such as:

(a) The decommissioning strategy selected;
(b) The time available;
(c) The availability of funds;
(d) Individual and collective doses to workers;
(e) Liquid and airborne discharges and their radiological impact on the general public and the environment;
(f) Industrial safety requirements;
(g) Available waste management and disposal options;
(h) Workforce availability, including contractors;
(i) Reuse of the buildings for other purposes.

Within established constraints, the optimal decision will in general be based on a multiattribute analysis or an extended cost–benefit analysis [28, 29].

The extensiveness of the decontamination will depend on the decommissioning strategy selected. In a delayed dismantling scenario, natural decay will reduce radiation and contamination levels in plant systems and components as well as on surfaces and may render some decontamination superfluous. When the need remains after a long SE time, the effect of physicochemical mechanisms during SE may make decontamination less effective, e.g. due to corrosion layers on metals and deeper migration into concrete surfaces.

If SE is planned, decontamination will be considered primarily for the areas that will be accessed during the transition period. An alternative in some cases may be to fix contamination in place to reduce airborne resuspension and facilitate access. However, it is important that surface coatings do not overly complicate future decontamination and measurement.

11.4.2. System decontamination

System decontamination may be performed on radioactive systems in order to reduce the general activity level within the systems in preparation for work during the transition period. System decontamination should be carried out while qualified personnel with knowledge of the relevant systems are still available.

Various decontamination methods are possible and it is important that the method and decontamination chemicals be chosen with a view towards available waste treatment installations and minimization of secondary waste. For example, a solution with a suitable composition and temperature for dissolution of the activity containing oxide layer on the surfaces of a system can be circulated in the system to transport the dissolved activity to a filter or ion exchange resin which is subsequently disposed of. After decontamination has been completed, the systems are flushed and dried.

Experience with system decontamination has been favourable for both typical BWR and PWR oxides [30]. In addition, more efficient methods with easier to handle final products are constantly being developed.

11.5. CHARACTERIZATION AND INVENTORY OF RADIOACTIVE AND HAZARDOUS MATERIALS

The characterization and establishment of an inventory of radioactive and hazardous materials within the facility involves surveys of existing data, calculations, in situ measurements and/or sampling and analysis. A database can then be established which will provide significant input into the decommissioning planning process and the development of successful implementation plans. With this database, management may assess and decide on various options and their consequences such as:

(a) Operating techniques: decontamination processes, dismantling procedures (hands-on, semi-remote or fully remote) and the required equipment;
(b) Radiological and industrial protection of the workers, the public and the environment;
(c) Waste management, waste classification and disposal options;

(d) Discharge authorization;
(e) Cost profiles.

At the beginning of the transition period, sufficient information should be collected to assess the radiological status of the facility and the nature and extent of any other hazardous materials present. Data collected during this initial characterization period would generally be based on information available at the time of final shutdown, including historical operating records. A survey of the extent of contaminated land should be made early in the transition period.

As work progresses during the transition period, the objectives of characterization move towards developing more detailed data concerning the physical, chemical and radiological conditions of the facility, including contaminated land. This will include activation calculations, taking and analyzing of samples, as well as in situ measurements of dose rates (Figs 8, 9) and contamination to fill the gaps in the available information.

Information gathered during these phases serves as the technical basis for work and project decisions, including cost estimates, exposure estimates, risk evaluation, waste management, scheduling and workforce requirements, particularly with respect to radiological exposures. Since characterization requires time, money and dose commitment, it should be optimized to meet the above objectives. Further guidance can be found in Ref. [21]. As this information should be updated on a regular basis, it is important that the database remain active during the entire decommissioning period.

11.6. CONSIDERATION OF SYSTEMS

Decisions on which systems must remain functional should be made during the planning of the transition and are based on:

(a) An evaluation to ensure that safety requirements will continue to be met,
(b) Support of human entry or occupancy for surveillance and maintenance,
(c) Possible use during future phases of decommissioning (see Section 9.2.),
(d) Restrictions posed by the current operating licence.

Some considerations may require cost–benefit analysis of:

(1) Energy consumption, surveillance and maintenance requirements;
(2) Replacement of complex systems with simpler ones;
(3) The possible need to achieve a safer state;

(4) A diminished need for redundancy;
(5) Operational and structural reliability over the anticipated duration of transition and SE;
(6) Demolition of buildings that contain systems or components which must be moved elsewhere.

During planning of the transition period, decisions regarding systems and major equipment within a facility may need to consider the following options:

(i) Operable as is: Systems that must remain operable and do not require modification (for example, lighting where surveillance and maintenance is to be done).
(ii) Modified: Some systems will need to remain operable but, as a result of the above assessments, modifications are required. For example, building ventilation is needed to maintain control of remaining contaminated areas but its design capacity is excessive, or redundancy of systems and components is no longer required because the consequences of temporary failure are acceptable until repairs can be made.
(iii) Preserved for future use: A limited number of systems and equipment may be preserved for the future. For example, installed manipulators and cranes can be of use during dismantling, or radioactive waste treatment systems may be valuable for processing decontamination solutions. Decisions of this type will depend on the length of time until such use is expected as some ageing will occur even in systems that are not in operation. Future refurbishment may be needed to bring these preserved systems and equipment to satisfactory levels of operability.
(iv) New: In some cases system functions will be needed, but use of the installed system may not be feasible because it may be overly complex, be over capacity, have high levels of contamination, or entail difficulty of access for operation or maintenance. In such cases, total replacement with new systems and/or equipment is the prudent course of action. Installation of a new ventilation system is a typical example [31]. Others include replacement of instrumentation because of obsolescence or the need for monitoring from a different or a remote location, and the installation of limited lighting for infrequent inspections where isolation of other unused circuits is not practical. A third example is a new electrical distribution system to repower that equipment necessary to support the decommissioning work. The design and installation of such a system in preparation for the decommissioning of the Big Rock NPP in the USA is described in Ref. [32].

(v) Retired: In many cases, a large number of systems will no longer be needed. In such situations, they are generally left in place and suitably isolated using standard safety practices, especially where there is internal radiological or hazardous chemical contamination or, in the case of electrical systems, the potential for short circuits or high voltage shocks. In some cases, complete removal of a system may be chosen, for example when the assets can be used at other facilities, or systems such as installed ventilation may be isolated where it is beneficial to use temporary or portable equipment when needed.

Once the decisions on systems have been made, the end point specifications and requirements for system surveillance and maintenance during the transition period and SE can be determined.

11.7. PREPARATION OF A FACILITY'S ROOMS AND BUILDINGS DURING TRANSITION

During the transition period, access to rooms and buildings in a facility needs to be defined in at least three ways: routine access, no access and completely isolated.

(a) Routine access: Human access for surveillance and/or maintenance can be as frequent as daily or as infrequent as, say, every three months. Industrial safety standards can be provided by either temporary, portable or permanent means. Ventilation, lighting and other safety measures are made available, although they are not necessarily in operation when the area is unoccupied. Walkthrough routes for periodic surveillance of unoccupied buildings are reviewed for industrial hazards and appropriate protection put in place (e.g. guardrails, warning signs, selected electrical isolators). Contamination and radiation zones will be tightly controlled and delineated to prevent the migration of contamination. Services normally found in continuously occupied facilities (e.g. toilets, drinking water) need not be provided.

(b) No access anticipated: Access will not be required, or if so the need will be so infrequent that special entry procedures can be established.

(c) Isolated: Entry will not be required until demolition begins.

Decisions as to the type of access needed to specific rooms and buildings are closely tied to an evaluation of the surveillance and maintenance requirements. When the surveillance and maintenance routines are determined and the access requirements are decided on, the results will be important inputs to creating the transition end point specifications. This process may include significant modifications to building access and other infrastructure in preparation for decommissioning. A detailed example is given in Ref. [33].

11.8. PROTECTION FROM EXTERNAL OR INTERNAL EVENTS

A number of external or internal events may affect a facility. For example, a fire prevention strategy is intended to eliminate fire hazards to the greatest possible extent. Some likely problem areas may include oils and grease in systems and components which, although emptied and flushed, may still contain residual material. Maintenance of good housekeeping standards and emergency access routes are key features in the implementation of such a strategy.

Flood protection may be a concern after shutdown, depending on the geographical location and the climate, geology and hydrology of the area. Some areas may require sealing or the maintenance of active collection, detection and pumping systems.

11.9. REMOVAL OF MINOR COMPONENTS

Generally, no major dismantling of radioactive parts of a plant takes place during the transition period, depending on the licensing regime. For example, under US regulations, major dismantling activities are defined as any activity that results in permanent removal of major radioactive components, permanently modifies the structure of the containment, or results in dismantling for shipment of components which contain greater than class C waste, i.e. waste unsuitable for routine near surface disposal. Major radioactive components defined by these regulations could include the reactor vessel and internals, steam generators, pressurizers, large bore reactor coolant system piping and other large components that are radioactive to a comparable degree [34].

Examples of decommissioning activities which are considered minor are:

(a) Normal maintenance and repair;
(b) Removal of certain, relatively small radioactive components such as control rod drive mechanisms, pumps, piping and valves;

(c) Removal of components (other than those defined above as major components) similar to those normally removed for maintenance and repair during plant operations;

(d) Removal of non-radioactive components and structures not required for safety. This can entail significant amounts of work and include major non-radioactive components such as cooling towers, transformers and control panels. Figure 13 gives an example of such activities at the Würgassen NPP, Germany.

During the transition period, removal of readily movable equipment which is no longer needed can be considered. These items are either:

(1) Packaged and disposed of;
(2) Packaged after compaction and disposed of;
(3) Decontaminated (e.g. by steam blaster or high pressure water jet);
(4) Directly released without treatment.

FIG. 13. Dismantling of cooling towers during the transition period at the Würgassen NPP, Germany.

12. COST OF TRANSITION ACTIVITIES

The costs of transition activities can be significant and a lack of timely funds during the transition period will severely impair the progress of the work. This section describes a number of activities to be carried out during the transition period. Decommissioning costs, including the costs of transition activities, are categorized in a proposed standardized list [35]. The list, with a focus on the transition period, is shown in the appendix and includes the following groups:

(a) Pre-decommissioning actions, e.g. decommissioning planning;
(b) Facility shutdown activities, e.g. removal of the spent fuel, system reconfiguration and retirement, decontamination and immobilization of residual contamination;
(c) (Limited) procurement of equipment and materials;
(d) (Limited) dismantling activities and characterization of radioactive inventory;
(e) Waste processing, storage and disposal (including hazardous waste);
(f) Site security, surveillance and maintenance;
(g) Transition project management;
(h) Other costs, including asset recovery.

The prime costs of the transition period activities are related to labour and fuel removal activities, but also include the purchase of equipment and consumables, contract work, etc. The costs are plant specific and dependent on whichever other activities are being pursued on the site. They are also dependent on the schedule chosen for shutdown of the plant and the start of decommissioning.

Input data for decommissioning cost estimates are available from international organizations [36] as well as commercially, for example the parametric cost estimating database of the United Kingdom Atomic Energy Authority (UKAEA) outlined in Annex I–13. Some of these systems have been developed for specific types of facility and should be used with caution for other types. However, their continued use and collaborative data sharing will improve their applicability across the range of nuclear facilities.

The costs for specific activities within the transition period should be clearly allocated to the operational or decommissioning base costs to establish an unambiguous boundary. Evaluating decommissioning cost according to a standardized list of cost items [35], including the costs of transition activities, would facilitate the comparison of costs for various decommissioning projects and the assessment of cost differences.

13. CONCLUSIONS

It is very desirable to take timely action to place a nuclear facility in a safe, stable and known condition as soon as possible after final shutdown. It is important that stabilization and other activities for facilities, systems and materials be planned and initiated prior to the end of operations. Carrying out these activities during the final stages of a facility's operational phase and during the transition period will be beneficial in that the operational capabilities of the facility and the knowledge of personnel will be utilized before they are lost. Actions taken at this time will pave the way to efficient and cost effective decommissioning by eliminating, reducing or mitigating hazards, minimizing uncertainty and maintaining steady progress.

The main conclusions of this report are that:

(a) Early planning is the key to a smooth transition from operation to decommissioning and will avoid a no action scenario.
(b) Planning for transition requires timely allocation of dedicated human, technical and financial resources.
(c) Timely implementation of transition activities will reduce expenditures and hazards, simplify waste and material management and help to keep the workforce motivated.
(d) Significant cultural and organizational changes will occur during the transition from operation to decommissioning and need appropriate consideration and management.
(e) The availability of relevant data and records is essential for smooth progress into and implementation of decommissioning. A database containing all relevant data needs to be established and maintained. This database should be kept up to date throughout the lifetime of the facility.
(f) Implementation of transition will require comparable management focus and workforce attention to detail as during normal operation.
(g) Good communication and involvement of all relevant stakeholders is essential for a successful transition from operation to decommissioning.

Appendix

STANDARDIZED COST ITEMS FOR DECOMMISSIONING PROJECTS [35][5]

01 PRE-DECOMMISSIONING ACTIONS

01.0100 Decommissioning planning

01.0200 Authorization

01.0300 Radiological surveys for planning and licensing

01.0400 Hazardous material surveys and analysis

01.0500 Prime contracting selection

02 FACILITY SHUTDOWN ACTIVITIES

02.0100 Plant shutdown and inspection

02.0200 Removal of fuel and/or nuclear materials

02.0300 Drainage and drying or blowdown of all systems not in operation

02.0400 Sampling for radiological inventory characterization after plant shutdown, defuelling and drainage and drying or blowdown of systems

02.0500 Removal of system fluids (water, oils, etc.)

02.0600 Removal of special system fluids (D_2O, sodium, etc.)

02.0700 Decontamination of systems for dose reduction

02.0800 Removal of waste from decontamination

02.0900 Removal of combustible material

02.1000 Removal of spent resins

---

[5] Actions relevant to the transition phase appear in italics; actions partly or possibly relevant to it are in normal typeface, and actions not relevant to transition appear in bold italics.

02.1100 Removal of other waste from facility operations

02.1200 Isolation of power equipment

02.1300 Asset recovery: resale/transfer of facility equipment and components as well as surplus inventory to other licensed (contaminated) and unlicensed (non-contaminated) facilities

03 PROCUREMENT OF GENERAL EQUIPMENT AND MATERIAL

03.0100 General site dismantling equipment

03.0200 General equipment for personnel/tooling decontamination

03.0300 General radiation protection and health physics equipment

03.0400 General security and maintenance equipment for long term storage

04 DISMANTLING ACTIVITIES

04.0100 Decontamination of areas and equipment in buildings to facilitate dismantling

04.0200 Drainage of spent fuel pool and decontamination of linings

04.0300 Preparation for dormancy

04.0400 Dismantling and transfer of contaminated equipment and material to the containment structure for long term storage

04.0500 Sampling for radiological inventory characterization in the installations after zoning and in view of dormancy

04.0600 Site reconfiguration, isolating and securing structures

04.0700 Facility (controlled area) hardening, isolation or entombment

04.0800 Radiological inventory characterization for decommissioning and decontamination

04.0900 Preparation of temporary waste storage area

04.1000 Removal of fuel handling equipment

04.1100 Design, procurement and testing of special tooling/equipment for remote dismantling

04.1200 Dismantling operations on reactor vessel and internals

04.1300 Removal of primary and auxiliary systems

04.1400 Removal of biological/thermal shield

04.1500 Removal of other material/equipment from the containment structure and all other facilities, or removal of entire contaminated facilities

04.1600 Removal and disposal of asbestos

04.1700 Removal of pool linings

04.1800 Building decontamination

04.1900 Environmental cleanup

04.2000 Final radioactivity survey

04.2100 Characterization of radioactive materials

04.2200 Decontamination for recycling and reuse

04.2300 Personnel training

04.2400 Asset recovery: Sale/transfer of metal or materials, and salvaged equipment or components for recycling or reuse

05 WASTE PROCESSING, STORAGE AND DISPOSAL

05.0100 Waste processing, storage and disposal safety analysis

05.0200 Waste transport feasibility studies

05.0300 Special permits, packaging and transport requirements

05.0400 Processing of system fluids (water, oils, etc.) from facility operations

05.0500 Processing of special system fluids (D_2O, sodium, etc.) from facility operations

05.0600 Processing of waste from decontamination during facility operations

05.0700 Processing of combustible material from facility operations

05.0800 Processing of spent resins from facility operations

05.0900 Processing of other nuclear and hazardous materials from facility operations

05.1000 Storage of waste from facility operations

05.1100 Disposal of waste from facility operations

05.1200 Processing of decommissioning waste

05.1300 Packaging of decommissioning waste

05.1400 Transport of decommissioning waste

05.1500 Storage of decommissioning waste

05.1600 Disposal of decommissioning waste

06 SITE SECURITY, SURVEILLANCE AND MAINTENANCE

06.0100 Site security operation and surveillance

06.0200 Inspection and maintenance of buildings and systems in operation

06.0300 Site upkeep

06.0400 Energy and water

06.0500 Periodic radiation and environmental survey

07 SITE RESTORATION, CLEANUP AND LANDSCAPING

07.0100 Demolition or restoration of buildings

07.0200 Final cleanup and landscaping

07.0300 Independent compliance verification with cleanup and/or site reuse standards

07.0400 Perpetual funding/surveillance for limited or restricted release of property

08 PROJECT MANAGEMENT, ENGINEERING AND SITE SUPPORT

08.0100 Mobilization and preparatory work

08.0200 Project management and engineering services

08.0300 Public relations

08.0400 Support services

08.0500 Health and safety

08.0600 Demobilization

09 RESEARCH AND DEVELOPMENT

09.0100 Research and development of decontamination, radiation measurement and dismantling processes, tools and equipment

09.0200 Simulation of complicated work on model

10 FUEL AND NUCLEAR MATERIAL

10.0100 Transfer of fuel or nuclear material from the facility or from temporary storage to intermediate storage

10.0200 Intermediate storage

10.0300 Dismantling/disposal of the temporary storage facility

10.0400 Preparation of transfer of fuel or nuclear material from intermediate storage to final disposition

10.0500 Dismantling/disposal of intermediate storage facility

11 OTHER COSTS

11.0100 Owner costs

11.0200 General, overall (not specific) consulting costs

11.0300 General, overall (not specific) regulatory fees, inspections, certifications, reviews, etc.

11.0400 Taxes

11.0500 Insurance

11.0600 Overheads and general administration

11.0700 Contingency

11.0800 Interest on borrowed money

11.0900 Asset recovery: Resale/transfer of general equipment and material.

REFERENCES

[1] INTERNATIONAL ATOMIC ENERGY AGENCY, Decommissioning Techniques for Research Reactors, Technical Reports Series No. 373, IAEA, Vienna (1994).

[2] INTERNATIONAL ATOMIC ENERGY AGENCY, State-of-the-Art Technology for Decontamination and Dismantling of Nuclear Facilities, Technical Reports Series No. 395, IAEA, Vienna (1999).

[3] INTERNATIONAL ATOMIC ENERGY AGENCY, The Decommissioning of WWER Type Nuclear Power Plants, IAEA-TECDOC-1133, Vienna (2000).

[4] INTERNATIONAL ATOMIC ENERGY AGENCY, Organization and Management for Decommissioning of Large Nuclear Facilities, Technical Reports Series No. 399, IAEA, Vienna (2000).

[5] INTERNATIONAL ATOMIC ENERGY AGENCY, Managing the Early Termination of Operation of Nuclear Power Plants, Safety Reports Series No. 31, IAEA, Vienna (2003).

[6] INTERNATIONAL ATOMIC ENERGY AGENCY, Safety Considerations in the Transition from Operation to Decommissioning of Nuclear Facilities (in press).

[7] INTERNATIONAL ATOMIC ENERGY AGENCY, Decommissioning of Nuclear Power Plants and Research Reactors, Safety Standards Series No. WS-G-2.1, IAEA, Vienna (1999).

[8] INTERNATIONAL ATOMIC ENERGY AGENCY, Decommissioning of Nuclear Fuel Cycle Facilities, Safety Standards Series No. WS-G-2.4, IAEA, Vienna (2001).

[9] INTERNATIONAL ATOMIC ENERGY AGENCY, Decommissioning of Medical, Industrial and Research Facilities, Safety Standards Series No. WS-G-2.2, IAEA, Vienna (1999).

[10] KIRKLAND, M.C., "Decommissioning is inevitable: planning sooner rather than later can have big payoffs for nuclear power plants and utilities", WM'01 (Proc. Conf. Tucson, 2001), WM Symposia, Inc., Tucson, AZ (2001).

[11] INTERNATIONAL ATOMIC ENERGY AGENCY, Record Keeping for the Decommissioning of Nuclear Facilities: Guidelines and Experience, Technical Reports Series No. 411, IAEA, Vienna (2002).

[12] SZILAGYI, A., BRIN, G., "Operations to disposition: The final steps of a facility's operational phase", WM'00 (Proc. Conf. Tucson, 2000), WM Symposia, Inc., Tucson, AZ (2000).

[13] SCHULTZ, J.W., Overcoming the grief of plant closure, Nucl. Eng. Int. (**August 1994)** 22—23.

[14] LUNDBY, J.E., Decommissioning of a Uranium Reprocessing Pilot Plant, Final Report of the Nordic Nuclear Safety Research Project KAN-2.2, Rep. TemaNord 1994:594, May 1994, Nordic Council of Ministers, Copenhagen (1994).

[15] SPOONER, K., "Decommissioning of a UK nuclear power station in a national park — An owners perspective", WM'01 (Proc. Conf. Tucson, 2001), WM Symposia, Inc., Tucson, AZ (2001).
[16] KAY, J.M., ELLIS, E.T., WILLIAMS, T.W., "The Closure of Trawsfynydd power station — Effects on staff and the local community and identifying a strategy for decommissioning", Nuclear Decommissioning (Proc. Int. Conf. London, 1995), Mechanical Engineering Publications Ltd, London (1995) 295–304.
[17] OECD NUCLEAR ENERGY AGENCY, Report on the SCNI Workshop on Nuclear Power Plant Transition from Operation into Decommissioning: Human Factors and Organisation Considerations, Rep. NEA/CSNI/R(99)17, OECD, Paris (2000).
[18] KACZMARSKY, M.M., ODELL, W.H., "The successful transition from plant operations to decommissioning of Maine Yankee", WM'00 (Proc. Conf. Tucson, 2000), WM Symposia, Inc., Tucson, AZ (2000).
[19] McGARVEY, D.S., "Shutting down style: Managing nuclear decommissioning transition risks", The Best of D&D... Creative, Comprehensive and Cost-Efficient (Proc. Top. Mtg. Chicago, 1996), American Nuclear Society, La Grange Park, IL (1996).
[20] STRATMANN, W., BAECHLER, M., "Review of some aspects of radiological interest during the establishment of the safe enclosure of the THTR 300 plant", Technologies for Gas Cooled Reactor Decommissioning, Fuel Storage and Waste Disposal (Proc. Technical Committee Mtg Jülich, 1997), IAEA-TECDOC-1043, Vienna (1998).
[21] INTERNATIONAL ATOMIC ENERGY AGENCY, Radiological Characterization of Shut Down Nuclear Reactors for Decommissioning Purposes, Technical Reports Series No. 389, IAEA, Vienna (1998).
[22] INTERNATIONAL ATOMIC ENERGY AGENCY, Factors for Formulating Strategies for Environmental Restoration, IAEA-TECDOC-1032, Vienna (1998).
[23] U of Illinois reactor in SAFSTOR because fuel has nowhere to go, Decommissioning & Decontamination Monitor (**July 21 1999**) 4–7.
[24] HILDEBRAND, J.E., Planning ahead: Preparing for the early retirement and decommissioning of Oyster Creek, Radwaste Magazine (**Nov/Dec 1998**) 31–34.
[25] BARENTS, M.S., ANDREWS, P.J., "Progress with stage 1 decommissioning of the steam generating heavy water reactor (SGHWR)", WM'95 (Proc. Conf. Tucson, 2000), WM Symposia, Inc., Tucson, AZ (1995).
[26] BUCK, S., YOUNG, M.P., COLQUHOUN, A., "Decommissioning impacts on waste management", Decommissioning '91 (Proc. Symp. Tucson, 1991), WM Symposia, Inc., Tucson, AZ (1991) 163–170.
[27] GANGAHAR, M., PHILLIPS, A., To POCO or not to POCO, that is the question, Nucl. Energy **38** 6 (1999) 339–341.
[28] GORDELIER, S.C., PASSANT, F.H., "Decommissioning of Nuclear Electric's gas-cooled reactors: The development of a new strategy", Decommissioning Policies for Nuclear Facilities (Proc. Int. Sem. Paris, 1991), OECD, Paris (1992) 337–351.

[29] NUCLEAR REGULATORY COMMISSION, Identification and Evaluation of Facilitation Techniques for Decommissioning Light Water Power Reactors, Rep. NUREG/CR-3587, USNRC, Washington, DC (1986).

[30] INTERNATIONAL ATOMIC ENERGY AGENCY, Decontamination of Water Cooled Reactors, Technical Reports Series No. 365, IAEA, Vienna (1994).

[31] CAMPBELL, J.B., "Decommissioning of the B203 plant at BNFL Sellafield", WM'98 (Proc. Symp. Tucson, 1998), WM Symposia, Inc., Tucson, AZ (1998).

[32] KOUTSOUKOS, A., et al., "Design and installation of a decommissioning power system at the Big Rock nuclear plant" (Proc. 1999 American Power Conf. Chicago, 1999), PennWell Corp. (1999).

[33] HOUGH, N.J., "Decommissioning of the uranium processing 'amber area' at Dounreay, Scotland" (Proc. Int. Decommissioning Symp. Knoxville, 2000), USDOE, Washington, DC (2000).

[34] NUCLEAR REGULATORY COMMISSION, Staff Responses to Frequently Asked Questions Concerning Decommissioning of Nuclear Power Reactors, Rep. NUREG-1628, US Govt Printing Office, Washington, DC (1998).

[35] OECD NUCLEAR ENERGY AGENCY, INTERNATIONAL ATOMIC ENERGY AGENCY, EUROPEAN COMMISSION, Nuclear Decommissioning — A Proposed Standardised List of Items for Costing Purposes, Interim Technical Document, OECD, Paris (1999).

[36] EUROPEAN COMMISSION, EC DB NET, European Commission Data Base Network, www.sckcen.be/eccdecommissioning/about/database/initdb.htm.

Annex I

THE TRANSITION FROM OPERATION TO DECOMMISSIONING OF NUCLEAR FACILITIES — NATIONAL EXPERIENCE

The examples provided below of organization and management schemes for the transition from operation to decommissioning range from national policies and programmes to detailed technical and organizational aspects at individual facilities. It is felt that both approaches are useful to provide practical guidance on how transition projects are planned and managed in various Member States. The examples given are not necessarily best practices nor do they necessarily reflect the views of the IAEA; rather, they reflect a wide variety of national legislations and policies, social and economic conditions, nuclear programmes and traditions. Although the information presented is not intended to be exhaustive, the reader is encouraged to evaluate the applicability of these schemes to a specific transition project.

I–1. THE TRANSITION PERIOD IN THE CZECH REPUBLIC

I–1.1. Introduction

This section describes the decommissioning of nuclear installations in the Czech Republic, with emphasis on the period of transition from operation to decommissioning. Except for one zero power reactor, no nuclear installation has been decommissioned in the Czech Republic. The operating organizations in the Czech Republic are now preparing for the future decommissioning of their nuclear installations in accordance with relatively new legislation on the peaceful utilization of nuclear energy and ionizing radiation.

I–1.2. Overview of installations

There are two NPPs and three nuclear research reactors in the Czech Republic. One research reactor has already been decommissioned. There are also other non-reactor nuclear installations in the Czech Republic, e.g. an irradiation facility, research laboratories, etc. The nuclear installations are described in Tables I–I and I–II.

Only those installations of significance from the point of view of decommissioning are mentioned. The process of decommissioning installations such as irradiation facilities is very simple and does not pose a significant problem.

TABLE I–I. INSTALLATIONS WITH NUCLEAR REACTORS IN THE CZECH REUPBLIC

Nuclear installation	Type of reactor	Operating organization	Year of startup	Year of shutdown	Status
Dukovany NPP	4 WWER 440/213	ČEZ a.s.	Unit 1: 1985 Units 2, 3: 1986 Unit 4: 1987	2025–2027	In operation
Temelín NPP	2 WWER 1000/320	ČEZ a.s.	Unit 1: 2001 Unit 2: 2002	2041–2042	In operation
Research reactor LVR-15	Tank reactor 10 MW(th)	NRI Řež	VVR-S: 1957 LVR-15: 1989	2018	In operation
Research reactor LR-0	Zero power reactor	NRI Řež	TR-0: 1972 LR-0: 1982	2010	In operation
Training reactor VR-1	Zero power reactor	CTU Prague	1990	2020 or later	In operation
Research reactor ŠR-0	Zero power reactor	ŠKODA Nuclear Machinery	1970	1989	Decommissioned (1997)

TABLE I–II. OTHER INSTALLATIONS

Installation	Operating organization	Year of startup	Year of shutdown	Status
Research facilities (Hot cell facility, etc.)	NRI Řež	1958–1965	2005–2025	In operation
High level waste storage facility	NRI Řež	1995	2045	In operation

I–1.3. Licensing requirements related to transition

The main legislation applicable to the transition phase is Law No. 18/1997 Digest. on the peaceful utilization of nuclear energy and ionizing radiation (the Atomic Law), and Law No. 13/2002 Digest. which amended the Atomic Law.

The Atomic Law sets out the basic regulations relating to the decommissioning arrangements of nuclear installations. Decree No. 196/1999 Coll. issued on 21 August 1999 by the State Office for Nuclear Safety (SONS) specifically regulates the decommissioning of nuclear installations or workplaces with significant and very significant ionizing radiation sources.

This regulation details the method and extent of the assurance of radiation protection at decommissioning installations and workplaces with significant[1] or very significant ionizing radiation sources.[2] It establishes the decommissioning method and stipulates the extent and the format of the required documentation. The regulation also includes requirements to make financial provision for the decommissioning of nuclear installations. Decommissioning of nuclear installations will be up to the operating organizations, whose legal responsibility is to create a financial reserve for this purpose. Decommissioning, according to Czech legislation, means those activities aimed at releasing nuclear installations (or workplaces with ionizing radiation sources) for use for other purposes following the end of operations, or exempting them from the requirements of the Atomic Law.

The Atomic Law identifies the need for a specific licence to cover decommissioning work. The issue of a licence for individual stages of decommissioning of a nuclear installation or workplace with a significant or very significant ionizing radiation source requires specific documents to be produced. These include the documents listed in item G of the Appendix to the Atomic Law, which are paraphrased below:

(1) Evidence of financial provision for decommissioning of the facility;
(2) A description of the impact of operation of a nuclear installation on the local area;
(3) A description of technical methods and arrangements for decommissioning;
(4) A decommissioning schedule/programme;
(5) Methodologies for dismantling, decontamination, conditioning, transport, storage and disposal of parts of the installation contaminated by radionuclides;
(6) The assumed radionuclide inventory and predicted discharges into the environment and radioactive waste volumes generated;
(7) Radioactive waste management methods, including disposal;

[1] A significant source is one which, if involved in a radiation incident, could lead to acute health effects. However, there is no danger of a radiation accident associated with the source.

[2] Very significant sources are those which might precipitate a radiation accident.

(8) Arrangements for radioactive waste management (including forecast limiting cases) for the decommissioning process (requires SONS approval);
(9) Safety analysis;
(10) Arrangements for the measurement and assessment of radiation doses to personnel and the workplace, and environmental contamination by radionuclides and ionizing radiation (requires SONS approval);
(11) An on-site emergency plan (requires SONS approval);
(12) Evidence of physical protection/security of the decommissioned installation.

I–1.4. Decommissioning strategy/options taken into account

The Czech Republic has no separate decommissioning authority; decommissioning licences are issued by SONS. A case-by-case system is used for decommissioning with the decay period also being dependant on the nature of the installation. Decommissioning of nuclear installations to green field status is not obligatory.

Article 3 of Decree No. 196/1999 Coll. establishes strategic options for the decommissioning of nuclear installations or workplaces with ionizing radiation sources, based on:

(a) The linkages between decommissioning activities and the ability to separate clear stages. These include:
    (1) Continuous decommissioning, where the decommissioning activities are carried out in one stage immediately after the end of operation,
    (2) Gradual/staged decommissioning, where decommissioning is done in several defined and distinct stages.
(b) The scope of essential decontamination, dismantling and demolition work, of environmental monitoring and the possible reuse of land influence which of the following options can be adopted:
    (1) Direct decommissioning, when from a radiation protection perspective it is not necessary to carry out decontamination, dismantling and demolition work. In such circumstances, neither ionizing radiation sources nor equipment contaminated by the ionizing radiation sources remain at the site, or other ionizing radiation sources are switched off and isolated to preclude restart.
    (2) In situ decontamination and disposal is acceptable when it is practical to carry out sufficient decontamination of equipment without dismantling.
    (3) Integrated decommissioning and dismantling where it is necessary to dismantle equipment prior to decontamination and final disposal.

(4) Physical containment of the decontaminated facility undergoing decommissioning. In this situation, equipment is left in place, confined by protective barriers to prevent radionuclide leakage into the environment until these have decayed to a level stipulated by special regulations.

I–1.5. Description of the transition period

The transition period from operation to decommissioning is not defined by the regulations, nor is the term ‘transition’ used in the field of decommissioning. However, some activities could be regarded as transitional, e.g.

(a) Defuelling,
(b) Management of spent fuel,
(c) Drainage of systems,
(d) Pre-dismantling decontamination,
(e) Preparation for SE or care and surveillance.

Section I–1.6 gives detailed information about decommissioning of the installations.

The following is paraphrased from Art. 5 of Decree No. 196/1999 Coll.:

(1) The end of operation of a nuclear installation (or a workplace) significantly affects the radiological protection requirements; consequently, subsequent work remains under the control of SONS. The end of operation shall be reflected in changes to documentation, mainly in the determination of the decommissioning strategy, the demarcation of the controlled zone, the definition of the monitoring programme and emergency plans.
(2) Except in the case of premature termination of operation forced by an accident, the operational period of a nuclear reactor ends with the removal of the fuel from the reactor. In the case of a workplace with sealed ionizing radiation sources the operational period ends with the removal of the radiation sources.
(3) The transfer of the workplace to another legal body (licensee) does not legally constitute the end of operation of that workplace.

I–1.6. Description of the transition period of nuclear installations in the Czech Republic

I–1.6.1. Nuclear power plants

The Dukovany NPP was constructed between 1974 and 1985, with a break between 1976 and 1978 due to a change in the type of reactor being built (type V213 instead of V230). The first unit went into operation in 1985, the second and third in 1986 and the last in 1987. The first reactor of the Temelín NPP is now in commercial operation. The second is in trial operation.

Two basic options are considered with respect to the decommissioning of Czech NPPs:

(1) Immediate decommissioning after the termination of operation. The decommissioning activities are carried out immediately after the end of operation. Defuelling is done immediately and the spent fuel is transferred to the at-reactor pool. The duration of the cooling period depends on the spent fuel parameters. The fuel is then transported to the spent fuel storage facility. After defuelling, the primary circuit is decontaminated. Dismantling of non-contaminated equipment and buildings is started immediately. Pre-dismantling decontamination, dismantling and post-dismantling decontamination of equipment for handling radioactive wastes are carried out in contaminated buildings. Following the final decontamination of the buildings, demolition is started.

(2) Deferred decommissioning. The decommissioning activities are postponed.

   (i) Safe enclosure: Immediately after defuelling, the primary circuit is decontaminated and dismantling of non-contaminated equipment and buildings is started. The following activities are carried out simultaneously throughout the NPP except on the nuclear island: pre-dismantling decontamination, dismantling of equipment, post-dismantling decontamination and related processing of generated radioactive waste, final decontamination and demolition of active buildings. The nuclear island is the only remaining area of the NPPs. After the period of (SE) the remaining buildings and systems will be decommissioned.

   (ii) Storage with surveillance: Immediately after defuelling, the primary circuit is decontaminated. No other decontamination or dismantling operations are carried out. The buildings are maintained to provide physical containment. The NPP buildings are closed and kept under surveillance (physical protection, checking and maintenance of containment barriers, etc.) for a period of about 50 years, after which decommissioning is started.

Safe enclosure is the preferred option for the decommissioning of Czech NPPs. In the case of the Dukovany and Temelín NPPs, this option is being pursued with the planned condition for SE being:

(a) Dukovany NPP: The reactors (vessels with internals) will remain in the biological shield and the reactor buildings will be adapted for about 50 years of SE. Four buildings will remain on the site.
(b) Temelín NPP: The reactors will remain in their containment buildings. Two buildings will remain on the site for about 50 years.

I–1.6.2. Research and training reactors

LVR-15 research reactor

The LVR-15 is a light water moderated and cooled tank nuclear reactor with forced cooling. The reactor was operated at a maximum power of 2 MW from 1957 until 1969 when the power was increased to 4 MW. Following a change in fuel type in 1974 it operated at a maximum power of 6 MW. In 1988 the reactor was reconstructed and recommenced operations in 1989 at a maximum power of 10 MW. The reconstruction comprised replacement of the reactor vessel, primary circuit, reactor control system and ventilation system.

Because of the design of the reactor, the reactor system and equipment are all contaminated. Immediate dismantling has been selected as the appropriate decommissioning strategy. There are, however, some steps that could be considered parts of the transition process:

(a) Planning of the decommissioning;
(b) Defuelling and removal of the beryllium reflector and reactor internals;
(c) Dismantling of parts below the reactor lid;
(d) Dismantling of research loops, probes, irradiation channels and rabbit systems;
(e) Dismantling of the upper reactor lid;
(f) Draining of systems;
(g) Surveying and mapping of radiological conditions;
(h) Processing of radioactive wastes;
(i) Preliminary decontamination (if needed).

The spent fuel and beryllium reflector will be stored in the at-reactor pool, then in a remote pool and finally it will be transferred to the storage pool in the high level waste and spent fuel storage facility (Fig. I–1). The final disposal method for the fuel (reprocessing or disposal in a future deep geological repository) has not yet been decided on.

The above mentioned actions will require about two years. Then proper decommissioning will start and take about three years. Decommissioning will not include the demolition of the reactor building, which will be used for other purposes.

LR-0 research reactor

LR-0 is an experimental light water zero power reactor used to establish the core neutron physics characteristics and shielding requirements of the WWER type reactor. The TR-0 reactor was commissioned in 1972 as a heavy water zero power reactor. It was used for research in support of the ILS-150 reactor installed at the A-1 NPP (Slovakia). The TR-0 reactor was operated until 1979. From 1979 to 1982 it was reconstructed as the LR-0 light water

FIG. I–1. Storage facility for spent fuel and high level waste (HLW) with container for transporting LVR-15 spent fuel.

reactor. The LR-0 reactor began operation in 1982. Its maximum power is 5 kW (Fig. I–2).

A strategy of immediate dismantling has been selected. Decommissioning is expected to be relatively simple because the reactor design makes contamination of the reactor equipment unlikely. The reactor will be dismantled immediately. The following steps can be considered to be part of a transition period:

(1) Planning of decommissioning,
(2) Defuelling (the slightly irradiated fuel will either be used for fabrication of new fuel or be disposed of in a future deep geological repository),
(3) Draining of systems,
(4) Surveying and mapping of radiological conditions,
(5) Processing of potential radioactive wastes,
(6) Decontamination (if needed).

Decommissioning does not include demolition of the reactor building, which will be utilized for other purposes. The above mentioned activities will

FIG. I–2. The LR-0 research reactor hall.

take about two years. Decommissioning will then begin and is expected to take about two years.

VR-1 training reactor

The VR-1 training reactor is a pool type light water reactor which uses enriched uranium fuel. Its rated power is 1000 W (thermal). The decommissioning process for the reactor will be similar to the decommissioning of the LR-0 research reactor. The decommissioning does not include demolition of the reactor building, which will be utilized for other purposes.

ŠR-0 reactor

The ŠR-0 research reactor was a pool type light water reactor that used enriched uranium fuel. In 1989 the ŠR-0 reactor was not operational. Refurbishment of the reactor vessel and the shielding was planned. However, in 1990 it was decided to decommission the reactor instead. The ŠR-0 reactor was completely decommissioned between 1992 and 1996. In 1997, its operating licence lapsed.

I–1.6.3. Other installations

Some installations and facilities operated by NRI Řež will be decommissioned in the future (e.g. a hot and semi-hot cell facility, radiochemical and nuclear material laboratories, a high level waste and spent fuel storage facility).

The selected strategy for these facilities is also one of immediate dismantling. The following steps can be considered parts of the transition period:

— Planning of decommissioning,
— Characterization of materials and contamination,
— Removal of radioactive and/or nuclear materials,
— Dismantling of unusable equipment,
— Draining of systems,
— Surveying and mapping of radiological conditions,
— Processing of radioactive wastes,
— Preliminary decontamination.

Decommissioning does not include the demolition of buildings, which will be utilized for other purposes.

I–1.7. Issues in planning for decommissioning

In the past it was not necessary for operating organizations of nuclear installations to prepare a preliminary decommissioning plan as is now required by legislation. Now the decommissioning planning is an important part not only of the operation, but also of the planning and construction of a nuclear installation.

When the preliminary decommissioning plans of the operational nuclear facilities were prepared, some required data were either unavailable or unknown. This mainly related to research facilities which had been built many years previously. There was a lack of data available to assess the amounts of material arising from decommissioning (including information regarding the composition of materials, the level of contamination, etc.). Thus it was necessary to collect the necessary data (by measurement of actual dimensions, from the operational history or even from the construction data), perform measurements and carry out the calculations. Of course, the continuous collection of data will be used to prepare and update the final decommissioning plans.

I–1.8. Conclusions

No nuclear facility (except a zero power reactor) has been decommissioned in the Czech Republic. The operating organizations of the other facilities are now preparing for the decommissioning of their nuclear installations.

The transition period from operation to decommissioning of nuclear installations is not defined by the regulations and the term ‘transition’ is not formally used in this context. However, some activities can be regarded as transition activities such as defuelling, management of spent fuel, decontamination, drainage of systems and preparation for SE or storage with surveillance.

Nevertheless these activities, which constitute the interface between the end of operation and the start of decommissioning activities, are very important and have a great impact on the safe and successful implementation of the decommissioning programme.

I–2. TRANSITION ACTIVITIES AT THE DR-2 RESEARCH REACTOR, DENMARK

I–2.1. Introduction to the DR-2 research reactor

The DR-2 research reactor was a 5 MW thermal light water moderated and cooled tank type reactor with MTR type fuel elements. The reactor started operation at full power on 26 August 1959. Until 31 October 1963 it operated with three shifts per day, five days per week. From 1 November 1963 until its final shutdown on 31 October 1975 its operation was reduced to one shift per day, five days per week. The total integrated power during its operation was 5488 MW(th)·d.

The reason for the shutdown of DR-2 in 1975 was that it was felt that the DR-3 research reactor at Risø, a 10 MW(th) heavy water moderated and cooled reactor, could cover all Danish needs for neutron beam experiments and reactor irradiation. However, since there was some doubt as to whether this was correct, it was decided that the reactor should be shut down in such a way that it could be restarted easily. It was also decided that the reactor should be kept in SE and that its dismantling should be postponed until the other nuclear facilities at Risø, in particular the DR-3, were to be dismantled.

I–2.2. The initial transition activities

After final shutdown of DR-2 its core was dismantled and placed on the storage rack in the reactor tank. Two months later the fuel elements were moved to the fuel storage pool of the DR-3 from where they were sent to the USA. The beryllium reflector elements were left in the grid plate. The shim-safety rods were placed on the storage rack in their guide tubes, while the regulation rod was left hanging in its extension rod in the tank.

After removal of the fuel, the primary circuit, including the DR-2 tank and the holdup tank, was drained, as was the secondary circuit. It was discussed whether the primary system should be dried, but it was decided that this was not necessary. To provide the necessary radiation shield at the top of the reactor tank once the water in the tank had been removed, a steel plate was placed on top of the reactor carrying a 5 cm thick layer of lead bricks.

In 1978/1979 it was decided to abandon the restart option for DR-2 and to use the reactor hall for chemical engineering experiments. Since this meant that the reactor hall had to be transformed into a clean area a number of additional measures were taken. The control rod drive mechanisms at the reactor top were removed and the thickness of the lead brick shield was increased to 10 cm. In addition, a 40 cm thick concrete shield plate was placed on top of the reactor and sealed to the concrete block. All beam and irradiation tubes were, where needed, provided with additional shielding and steel plates were welded over the tube openings. Other openings to the reactor were sealed with a plastic material. The interior of the primary circuit and the tank were connected to the outside atmosphere through a filter. The staircase to the top of the reactor was removed and so were most railings. The secondary circuit was dismantled.

For the next 20 years the reactor was left in SE with little need for maintenance. Regular inspections and yearly radiation surveys were carried out. They indicated that the activity of the tank was decaying with a half-life of seven to eight years. Measurements of the γ spectrum revealed that, as expected, ^{60}Co was the dominant activity, but ^{152}Eu could also be detected.

I–2.3. The DR-2 project

The DR-2 project was started in October 1997. The aim of the project was:

(a) To determine the remaining activity levels in DR-2,
(b) To plan the final dismantling of DR-2.

An important reason for starting the project at that time was that staff at Risø with knowledge of DR-2 would soon be retiring and it was considered important that their knowledge be used while it was still available. The project planned to open the various parts of the reactor and to assess the remaining activity level, the radionuclides involved and where the activity was placed. However, before the reactor could be opened a number of activities had to be carried out. The reactor hall had been used for chemical engineering experiments between 1976 and 1996 and much equipment from these had been left and had to be removed before work could start. The walls and floor had to be cleaned and repainted, the staircase to the top of the reactor had to be reinstated, the reactor hall crane had to be re-licensed, etc.

To permit handling and measurement of radioactive components stored in the reactor, two facilities were built in the reactor hall. One, made from concrete blocks, was for the storage of radioactive components taken from the reactor. The other was a measuring facility, again made from concrete blocks and lead bricks, in which long components placed on a flat vehicle could be moved past a lead collimator where the activation distribution along the component could be measured.

In addition, since the status of the reactor had to be changed (from a sealed to an open facility) new safety documentation for the reactor had to be prepared, submitted to and approved by the nuclear and radiation safety authorities. The new documentation included a safety assessment of the project activities. To obtain the necessary information on the reactor the DR-2 archive, in particular drawings, had to be brought up to date. Permission to carry out the project came from the authorities in December 1999 but due to a leak in the DR-3 tank the opening of the reactor tank was postponed until May 1999.

I–2.4. Opening of the reactor tank

The initial task was the removal of the concrete shield plate (Fig. I–3) and the two layers of lead bricks. Then, through a hole in the steel plate, samples were taken of the air from the reactor tank and measured for activity and beryllium. No significant contamination was found. Next smear tests were carried out at the tank wall and at the surface of the reflector elements and they were tested for activity and beryllium. Again the levels detected were close to background. During these measurements the personnel on the reactor top carried respiratory protection. Based on the measurements performed this was not considered necessary during the subsequent work except when the

FIG. I–3. Twenty-five years after shutdown the concrete lid is lifted from the top of the DR-2 reactor, Denmark.

beryllium reflector elements were being removed from the reactor. The steel lid was taken away and the removal of components in the reactor tank could begin.

Initially it had been the plan to take movable components out of the reactor, measure their activity and put them back into the reactor. However, since this procedure would have to be repeated during the later dismantling process, it was decided to cut up the active components — when this could be done — and store them in waste drums which were, when filled up, transferred to the Risø waste treatment plant. Non-radioactive components were stored in plastic bags in the basement of the DR-2 building. Special stainless steel containers were made for the storage of the beryllium reflector elements.

The measurements involved determination of the total activity of the components by use of a γ spectrometer which also identified the radionuclides involved (the activity distribution along the components was also measured). The components taken out of the reactor tank were:

(a) Shim-safety rods, guide tubes and grid plate plugs from the storage rack;
(b) Reflector elements and 'water holes' from the grid plate (Fig. I–4);

FIG. I–4. The last movable part, a beryllium reflector element, is lifted out of the reactor tank of the DR-2 reactor, Denmark.

(c) The regulating rod, magnet rods for the safety rod and various other rods hanging from the top of the tank.

The activities of the more active components are given in Table I–III. It is seen from the table that the total activity of the movable components of the reactor tank was approximately 38 GBq or about 1 Ci, and that the activity is dominated by that of the regulating rod.

The dominant radionuclide was ^{60}Co. Only in the case of the Be reflector elements did the measurements indicate that 5–10% of the activity was ^{137}Cs.

When the lids were removed the maximum radiation level at the top of the reactor tank was 450 μSv/h. After all the components mentioned in Table I–III had been removed the radiation level was reduced to 75 μSv/h.

After the removal of all movable components from the reactor tank the radiation field in the tank was surveyed. This was done by use of thermoluminescent dosimeters. Close to the core region the radiation level was 25–50 mSv/h. It was noted that the radiation level was higher on the side of the thermal column than on the other side of the core. This is presumably due to the activity of the thermal column graphite (see Section I–2.6). It had not been possible to consider it separately but the grid plate with its stainless steel bolts and guide pins made a significant contribution to the radiation level close to the core region.

The compenents were removed without difficulty. Only one of the water hole plugs stuck and a hoist had to be used for its removal. There was no indication of corrosion.

TABLE I–III. ACTIVITY OF COMPONENTS REMOVED FROM THE DR-2 TANK

Components	Total activity (MBq)
4 safety rods (Al + B_4C + SS + Pb)	415
5 guide tubes (SS)	3279
1 safety rod and guide tube	919
3 grid plate plugs with SS screws	129
12 Be reflector elements (Be + SS)	10 000
1 regulation rod (SS)	23 000
Water holes, etc.	69
Total	37 800

I–2.5. Opening of the hold-up tank room

The hold-up tank room is situated directly under the reactor and was closed by a wall of concrete blocks. Some of these blocks were removed to permit access to the room. Five stainless steel boxes with components from the decommissioning of Risø's hot cell facility had previously been stored in the room. These were transferred to the waste management plant. There were also two heavy radioactive components, possibly a beam hole shutter and a magnet, which had undoubtedly been used in connection with reactor experiments. The shutter was moved to the reactor hall's storage facility while the magnet remained in the hold-up tank room. A number of loose non-radioactive components were also removed. After cleanup of the hold-up tank room the concrete block wall was re-established.

I–2.6. Opening of the igloo in front of the thermal column

The igloo in front of the thermal column had been used for storage of a number of reactor components such as ion chambers, old beam plugs, the tank television camera, vertical irradiation tubes, graphite stringers and other components. Their storage had been ad hoc in nature. Most of these items were activated to varying degrees. When one of the concrete blocks in front of the igloo was removed the radiation level was measured at 25–35 μSv/h at the opening (Fig. I–5).

The radioactive components that could be were cut, loaded into drums and sent to the waste treatment plant. An exception was the graphite stringers which were wrapped in plastic sheeting for later transport to the waste treatment plant. They may have to be annealed before final disposal. Components that were too heavy to be cut were stored in the storage facility in the reactor hall. Non-radioactive components were stored in plastic bags in the basement. After the igloo had been emptied, the radiation level at the entrance had been reduced to less than 1 μSv/h.

Next the other concrete block in front of the igloo was moved and the motor driven shield door was rolled out. This provided free access to the outer surface of the thermal column. A number of graphite stringers, both outer and inner, were taken out of the column, measured and reinserted. It was a surprise that the stringers contained ^{152}Eu, ^{154}Eu and ^{60}Co activity, of which the ^{152}Eu activity was dominant. It has been estimated that the total activity of the graphite amounts to about 5 GBq. Also unexpectedly the activity of the stringers generally decayed out through the column exponentially. Most of both the inner and the outer stringers showed maximum activity around their middles. This seems to indicate that the europium in the graphite stringers is

FIG. I–5. The igloo in front of the thermal column just after one of the concrete blocks in front of the igloo was removed (DR-2 reactor, Denmark).

not evenly distributed. Measurements of the activity of a thermocouple taken from the thermal column gave the expected exponential decay pattern. When both stringers were out, lead samples were drilled out of the lead plate at the inner end of the thermal column and their activity was measured.

I–2.7. Opening of beam, irradiation and instrument tubes

Two beam plugs were taken out and their activity measured. The removable inner beam hole liner was also taken out of one and measured. While the liner was out the radiation level through the concrete shield was measured. To reduce the radiation from inside the tank, a container with lead bricks was lowered into the tank from the top to close the hole. The through tube, which had been used as a fast pneumatic facility, was also opened and the pneumatic tubing taken out and cut up. After the measurements the liner and the beam plugs were reinserted. An approximate estimate of the total activity of the eight beam tubes suggested an activity of the order of one GBq.

Some difficulties were experienced in extracting the beam plugs and the liners. The liner extraction was successful in only one of the two cases. An attempt to extract the largest beam tube was abandoned. However, it is believed that all the plugs and liners can be extracted by use of sufficient force when the need to reinsert the beam plugs and liners no longer exists. Two irradiation or S tubes were also taken out, measured and reinserted. Based on the activity measurements the total activity of the six S tubes is of the order of 0.1 GBq. The instrument tubes were opened and the radiation level through the tubes was measured.

I–2.8. Drilling of cored holes through the concrete shielding

Two holes were drilled through the concrete shield, the lead shield around the reactor tank and the reactor tank wall. The cores were extracted and measured (Fig. I–6). The results showed only the inner 50–60 cm of the concrete shield to have activity above the background level.

The concrete shield of the DR-2 is provided with a number of vertical tubes at different distances from the tank wall. These tubes were used to measure the activation of the concrete shield. Such measurements have been made using TL dosimeters.

I–2.9. Planning the final dismantling of DR-2

In September 2000 it was decided to decommission all the nuclear facilities at Risø. Therefore, planning for the final dismantling of DR-2 will be

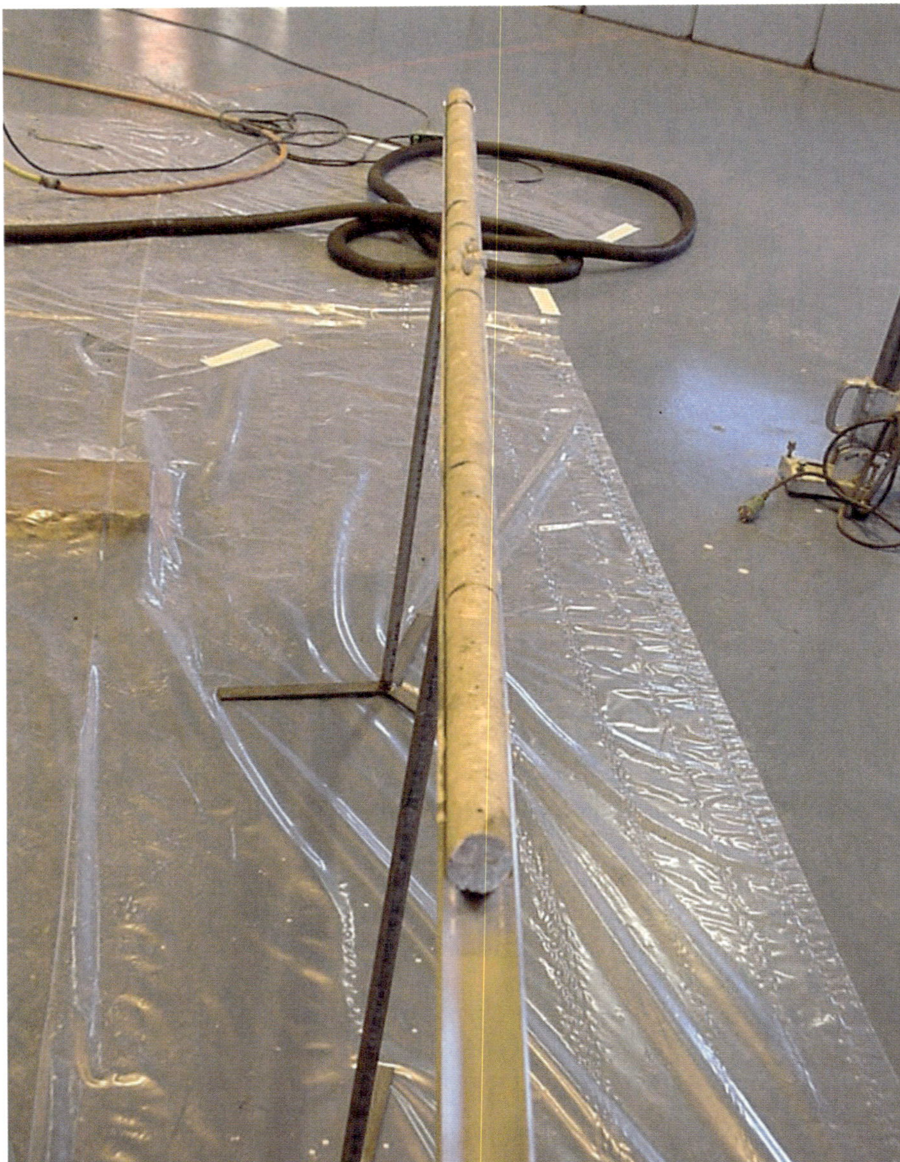

FIG. I–6. The borehole core resulting from drilling through the concrete shield of the DR-2 reactor, Denmark.

part of the overall Risø decommissioning programme and not part of a specific DR-2 project.

I–2.10. Conclusions

The DR-2 project is about to be terminated. It has been carried out with few difficulties according to the original plan. There have been a few surprises, but none that have caused significant difficulties. In general the difficulties have turned out to be less severe than expected. From a dose point of view it has been a significant advantage that the DR-2 has been in SE for more than 25 years.

I–3. TRANSITION TO DECOMMISSIONING: THE AT1 CASE, FRANCE

I–3.1. Introduction

France's nuclear decommissioning activities involve both the civilian and defence sectors. France has four major civilian operating organizations:

(1) Electricité de France (EDF),
(2) Compagnie Générale des Matières Nucléaires (Cogéma),
(3) Commissariat à l'Énergie Atomique (CEA),
(4) Agence Nationale pour les Déchets Radioactifs (ANDRA).

Nuclear energy supplies France with nearly 80% of its electricity. Presently EDF has 58 operating NPPs (PWRs). The six older gas cooled, graphite moderated NPPs have been shut down, as has Superphénix, a fast breeder reactor. Cogéma operates fuel cycle plants, including chemical reprocessing facilities, uranium production facilities and gaseous diffusion plants. Most of the nuclear R&D installations in France belong to the CEA. Many have already been dismantled or are presently being dismantled. Others (research reactors, laboratories, pilot plant) are awaiting decommissioning.

ANDRA was established in 1991 to take charge of the development and operation of waste disposal centres for low level waste (LLW), medium level waste (MLW), HLW and very low level waste (VLLW)). The first depository for waste containing short lived radionuclides, the CM Centre de la Manche, closed in 1994 after 25 years of operation and disposal of more than 530 000 m^3 of waste. The Aube centre (CA) came into operation in 1991 and has a capacity of 1 000 000 m^3. ANDRA plans to open a disposal facility for VLLW (from 0 to a few hundreds of Bq/g) close to the Aube centre in the near future.

I–3.2. National policy

I–3.2.1. Regulation

The basic nuclear installations (BNI) associated solely with the nuclear deterrent programme are regulated by a body under the authority of the CEA's high commissioner. The operating organization has to comply with the sixth decree of 11 December 1963 which sets out the requirements for decommissioning of BNIs. This decree established the Nuclear Safety Authority (DGSNR) (new name for DSIN). The sixth article, issued on 19 January 1990, modified the procedure for decommissioning BNIs. Before this article there were no requirements applying specifically to decommissioning. The applicable procedures were the general ones relating to major modifications of installations. An administrative note (9 November 1990) was issued to explain the new procedure in detail.

In regulatory terms, the decommissioning of nuclear facilities now requires three major phases. The first phase leads to the end of operation of the facility. Work is carried out in accordance with existing operating procedures. This phase includes the removal of all fuel, fertile and fissile materials, all waste produced during the operational phase and still present on the site, and some decontamination and drainage activities. The operating organization must inform the DGSNR of its intent to commence these operations six months prior to starting them and submit a safety case for the work. The DGSNR formally acknowledges the end of this phase following receipt of a completion report and after a thorough site inspection.

The second phase leads to shutdown of the facility. This phase, which can be started while the previous one is ongoing, consists of dismantling any equipment outside the nuclear island which is no longer required for surveillance and safety. The containment is also reinforced. At the end of this phase a complete inventory of the remaining radioactivity is produced. These operations require formal authorization from the government. This is only given after thorough examination of the necessary documents (safety analysis report, proposed general operating rules, decommissioning plan, etc.) by the DGSNR. The first two phases constitute the transition from operation to decommissioning. The third phase consists of dismantling the facility. This phase can be started as soon as the previous one is completed or it can be delayed. Authorization and assessment by the safety experts are needed prior to commencement.

The licensee's nuclear responsibilities for the site ends when all the radioactive materials are removed and this has been confirmed by the DGSNR on the basis of a completion report and a thorough site inspection. The procedure and associated arrangements are currently being reviewed by the DGSNR following the experience gained in recent years.

I–3.2.2. Strategy

When a BNI ceases operation some decontamination and modification activities are carried out prior to decommissioning, which is then followed by dismantling. This work will result, from the administrative standpoint, in the creation of a new BNI, in the reclassification of the BNI as an installation requiring licensing or registration solely on the grounds of environmental protection, or simply in a return to the public sector. The actual classification will depend on the remaining radioactivity levels within the facility.

Experience gained during previous dismantling operations, mainly of small installations (pilot installations, research reactors) led, in 1990, to regulations setting out the requirements for licence termination for BNIs. Currently, the operating organization is required to carefully consider the final condition of the installation and to organize the decommissioning and dismantling in several stages. The purpose of these measures is to ensure that the installation will remain in a satisfactory safety condition at all times, even after it has ceased to operate, and taking into account the nature of the dismantling operations.

The DGSNR considers the dismantling operations currently proceeding as test cases, providing an opportunity for the operating organizations to define and implement, on the one hand, a dismantling strategy and, on the other hand, a management policy for the large amounts of radioactive waste which will be generated. If carried through to their conclusion they would also constitute examples demonstrating the technical and financial feasibility of an entire dismantling process.

As specified in the sixth article of the above mentioned decree of 11 December 1963, when an operating organization decides to close down its installation for any reason it must inform the Director of the DGSNR by sending him:

(a) A document justifying the configuration in which the installation will be left after final shutdown and indicating the various stages of subsequent dismantling;
(b) A safety analysis covering the final shutdown procedures and indicating subsequent plant safety provisions;

(c) The general surveillance and maintenance arrangements to ensure a satisfactory safety condition;
(d) An updated on-site emergency plan for the installation.

The current environmental protection regulations also require the operating organization to submit an environmental impact analysis pertaining to the proposals. The implementation of these various requirements is subject to approval by decree, signed by the government after consultation with the Interministerial Commission for Basic Nuclear Installations. In some cases, operations such as the unloading and removal of nuclear material, the disposal of fluids, decontamination and drainage operations can be performed under the operational authorization. This requires that previously imposed requirements, the safety analysis report and general operating rules currently in force, subject to certain modifications if necessary, can all be complied with. Where this is not possible such operations come under the provisions of a decommissioning decree.

From the regulatory standpoint, after the end of the operational phase two successive sets of activities have to be carried out:

(1) Final shutdown work, which covers all of the transition phase authorized by the decree. This work mainly entails the dismantling of equipment not on the nuclear island and not required for surveillance and safety, the preservation or reinforcement of the containment barriers and taking of a radioactivity inventory. In most cases, the transition phase is then complete;
(2) Dismantling work on the nuclear part of the plant. This work can start as soon as the final shutdown operations are completed or can be delayed to benefit from radioactive decay in activated or contaminated materials. These operations can lead to partial or total dismantling, depending on the ultimate condition selected.

Frequently, as soon as the nature of the installation — although it is still a BNI — is changed by the dismantling operations it is considered to be a new nuclear installation. Consequently, a new authorization is required involving the procedure previously described, including a public inquiry. In most cases such plants become storage facilities for their own internal equipment.

If dismantling work reaches the stage where the total radioactivity of the remaining radioactive substances is below the minimum level necessitating classification as a BNI, the plant can be removed from the list of BNIs, i.e. 'declassified'. Depending on the residual radioactivity level it could come under the provisions of the law which requires installations to be registered or licensed on environmental protection grounds.

The DGSNR will soon issue an updated directive concerning the various technical and administrative aspects of BNI decommissioning and dismantling. This document will notably take into account experience acquired in such operations since January 1990 when the previously mentioned decree of 11 December 1963 on nuclear installations was supplemented.

Restrictions on the use of buildings or land pertaining to a BNI being released from nuclear status will be considered in this revised directive. It now appears that, on the principle of applying caution, some restrictions such as radioactivity measurement during earth moving or unusual operations shall be imposed on future landowners.

If the safety criteria are completed, utilities are free to choose their decommissioning strategy and the techniques to be used. The decommissioning strategy chosen in the late 1980s by EDF consists of delaying the total dismantling 50 years to benefit from the ^{60}Co decay. The first and second phases are completed immediately after final shutdown and the third phase is further divided into two subphases. The first subphase is started at the end of phase two. The second subphase is delayed 50 years. In the interim, the installation is used to store the equipment left in place and is kept under surveillance (SE concept). After this period, the installation is fully dismantled. At the end of the dismantling operation, if there are no longer any radioactive materials on the site, it can be released for other purposes and is no longer a nuclear site.

Experience has shown that interim situations tend to become the norm. However, it is necessary to balance the benefits and drawbacks of delaying the shutdown and dismantling phases. In this respect in 1996 the DGSNR requested that the CEA and EDF produce a joint study assessing the possibility of shortening the planned 50 year delay at the EL4 HWGCR located in Brittany. The CEA and EDF proposed, as a result of their review, to proceed to total decommissioning immediately after the end of operations.

In April 2001 EDF took the decision to proceed with immediate decommissioning of all the GCRs, the first PWR (CHOOZ A), EL4 and Superphénix. The aims of this new strategy are:

(i) To show the feasibility of total dismantling on an industrial scale,
(ii) To show the capability of managing all the generated material and wastes,
(iii) To take the opportunity of gaining internal decommissioning expertise,
(iv) To be able to take charge of the future decommissioning of the PWRs.

CEA's decommissioning strategy for its R&D installations is to:

(a) Start decontamination and decommissioning studies before the shutdown decision is taken;
(b) Carry out operations as soon as possible after shutdown, while using the operating staff to:
    - Benefit from experienced operating staff,
    - Deal with cleaning and cleansing problems,
    - Minimize operating costs,
    - Optimize the safety classification;
(c) Spread or postpone partial or total DECOM (demolishing of civil works) within the constraints of available financing and waste shipments to existing depositories.

I–3.3. The AT1 case

The AT1 pilot facility on the Cogéma site at La Hague near Cherbourg was built to reprocess spent fuel elements from fast breeder reactors (Rapsodie and Phénix). It was operated for ten years, from 1969 to 1979, during which time it produced 1094 kg of uranium and plutonium at a rate of 2 kg per day. Initially the facility implemented only three extraction cycles; a fourth uranium/plutonium extraction cycle was added in 1972 together with an additional fission product storage unit in the extension wing. After final shutdown in 1979 the process lines were rinsed for nearly three years, during which 600 kg of plutonium and 1700 kg of uranium were recovered. All the process lines were systematically decontaminated between July 1979 and December 1981. This transition period was carried out in five main stages:

(1) Rinsing and cleaning of equipment, circuits and tanks, to recover fissile materials.
(2) Decontamination of the fission product tanks.
(3) Dismantling of in-cell equipment: where possible equipment was removed, after remote decontamination, in concrete containers and shielded transport casks.
(4) Treatment of contaminated solvent: 1800 litres of spent solvent, containing up to 66 mg/L of plutonium and 480 mg/L of U, were treated to reduce the plutonium level to 1 mg/L and beta gamma contamination to less than 200 MBq/L. Decontaminated solvent was then transferred to Marcoule to be vitrified.
(5) Transfer of fission product bearing solutions. Two tanks, containing 7 m^3 and 13 m^3, respectively, were emptied to the central fission product storage unit where the solution was vitrified.

The CEA's decommissioning unit for nuclear facilities (DDCO) took over the AT1 decommissioning programme in 1982. It was to carry out the total decommissioning of the installation with the exception of the civil engineering. This involved the following operations:

(a) Dismantling and removal of all contaminated process lines and equipment,
(b) Decontamination of the process cells to the lowest possible residual level to allow the building to be reused without requiring continuous radiological monitoring.

The process functions carried out in the AT1 facility are shown in Fig. I–7.

Fuel elements in lead transport casks entered cell 901 from which they were transferred to cell 902. Here they were cut up with hydraulic shears and placed in baskets. The baskets were then lowered from cell 902 into the dissolver in cell 903. Following dissolution, the hot nitric acid solution from the dissolver was clarified and sent to the three extraction cycles in cells 904 and 905. After extraction of the fission products, the fissionable materials were routed to the end of the process to separate the uranium and plutonium, concentrate them and convert them into uranyl nitrate and plutonium oxide. The equipment required for these processes was located in the alpha cells and unshielded gloveboxes (cells 950–952 of the extension wing, and cell 906). Other cells were used for fission product storage (cells 908 and 909 in the main building and cell 920 in the extension wing), liquid waste storage (cell 907), off-gas treatment (filtration cell) and gloveboxes for analysis sampling.

I–3.4. Implementation

Rinsing, cleaning and decontamination of all the circuits and tanks took 30 months after shutdown. During the transition phase, AT1 was operated as during the operational phase:

(a) During the rinsing phase, five shifts worked for 8 h each;
(b) During the decontamination phase, three 8 h shifts were worked (excluding Saturday and Sunday).

Thirty people were employed in the operation (including management) and were supported by 20 people carrying out laboratory analyses. Dose intake was the same as during operation. During this period, when AT1 was operated normally, it was necessary to make small investments such as in high pressure pumps and shielded gloveboxes (treatment of contaminated solvent). It was also necessary to create a new facility to allow the transfer of the liquid fission product solution. All rinsing and cleaning were done with progressive addition of hydrofluoric acid and were conducted through the normal AT1 process and circuits. This allowed the alpha contaminated waste volumes to be minimized. The residual surface alpha contamination was between 5 and 70 MBq/m^2. The decontamination factor was between 3500 and 250. This was achieved by using a rinsing volume five times higher than the total volume of the equipment and circuits. More than 600 g of plutonium and 1700 g of uranium, double the

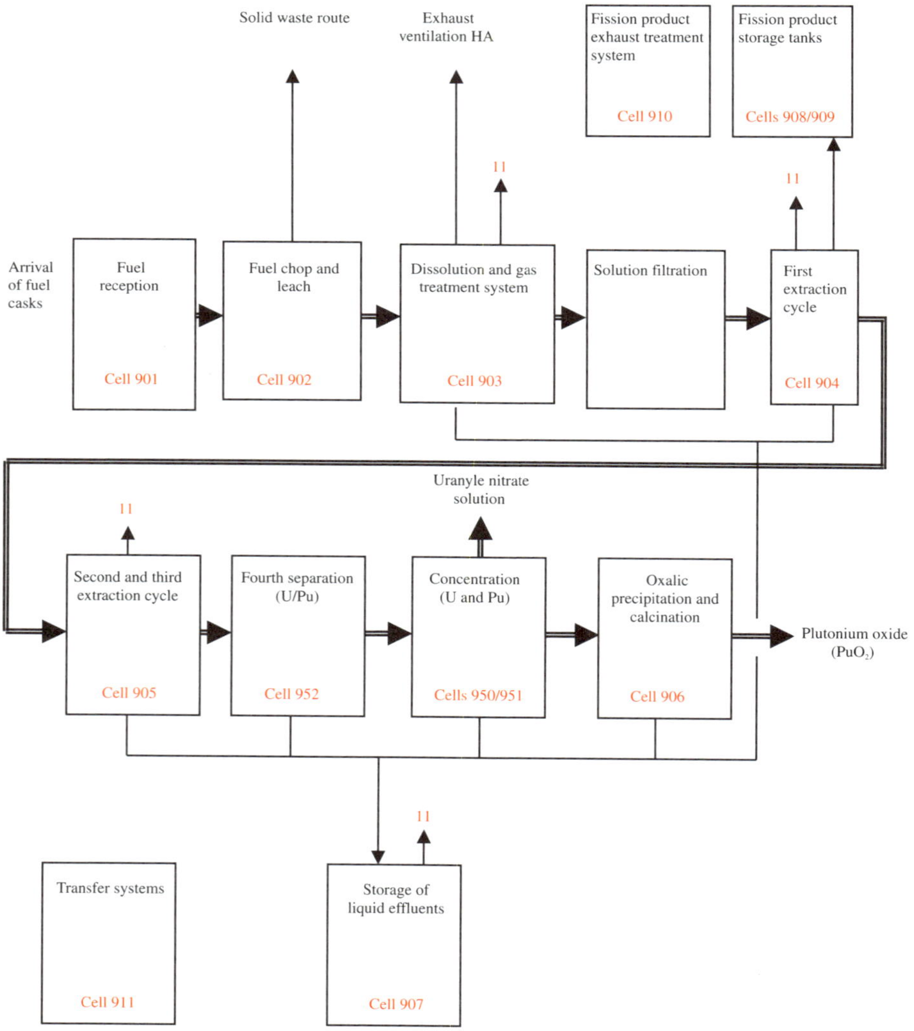

FIG. I–7. Process schematic (with functional locations) of the AT1 facility.

estimated quantities, were recovered through the rinsing operation. The rinsing operations also reduced beta gamma levels by a factor of 2 to 7. Decontamination was done using acid and basic chemicals (oxidation/reduction) through the process line and gave good results. Fission product storage tanks were decontaminated and refilled to allow further contact dismantling.

I–3.5. Conclusions

(a) The work in mechanical cells (901–903, filtration) shows that it is possible to reach an activity level consistent with dismantling if the equipment and systems have been designed with dismantling in mind.
(b) Treatment of spent solvent is possible but special attention has to be given to precipitate products and other residues.
(c) Decontamination of the concrete structures in cells 903 and 905 was made more difficult because of the use, during the transition phase, of high pressure steam and water. These processes push contamination into the concrete and facilitate contamination migration in the structure.
(d) During the transition phase, it is essential to obtain as much information about the facility's operational lifetime and records as possible. Former operating staff and the decommissioning team have to work together to avoid removal of equipment that might be used in the future and to avoid using processes which could force contamination into the concrete structure.

I–4. TRANSITION TO DECOMMISSIONING: THE UP1 CASE, FRANCE

I–4.1. Introduction

A major decommissioning project is under way at the UP1 reprocessing plant at Marcoule, southern France. This section provides an overview of the site and the plant itself, before discussing the organization of the project, its financing and objectives. In conclusion a more detailed update of the final shutdown programme that began in January 1998 is included.

I–4.2. Forty years of production at Cogéma/Marcoule

In 1952, the CEA selected Marcoule, located along the Rhône River 100 kilometres from the Mediterranean, as the first industrial site for developing its defence related nuclear activities. Marcoule has diversified its activities considerably in the last half century.

The site covers 280 hectares (nearly 700 acres) in a zone measuring about 1700 meters by 1600 meters and hosts a wide range of nuclear activities employing a total of about 6700 people. The site includes:

- Several reactors,
- A MOX fuel fabrication plant,
- Numerous laboratories,
- A low level waste incineration and melting facility,
- A reprocessing plant,
- Solid waste processing and interim storage facilities,
- Liquid waste treatment facilities,
- A fission product vitrification facility.

The last four facilities lie within the scope of the CODEM joint venture. The UP1 plant was commissioned in 1958 to reprocess spent fuel from the reactors used to produce plutonium for French defence. In 1976, with the creation of Cogéma, which took over the industrial operation of the site, UP1 began reprocessing spent fuel from commercial natural uranium fuelled gas cooled power reactors. These reactors were gradually phased out of production and after reprocessing 18 600 metric tons of spent nuclear fuel the UP1 plant itself was shut down at the end of 1997. The programme of decommissioning the facilities and disposing of the radioactive waste was initiated at the beginning of 1998. During its forty year service life, the UP1 plant was operated for the CEA (including the Ministry of Defence), for EDF and for Cogéma on behalf of its own clients.

I–4.3. Creation and mission of CODEM

In 1995 the French Ministries of Defence and Industry instructed three entities, CEA, EDF and Cogéma, to set up a 'customer structure' to manage the decommissioning operations following the shutdown of UP1. A joint venture known as CODEM was created on 1 July 1996, with 45% owned by CEA (including 40% on behalf of the Ministry of Defence), 45% by EDF and 10% by Cogéma on behalf of all the other clients served by the plant.

CODEM is the decision making, supervisory and funding entity for the decommissioning operations. Its three primary objectives are to maintain compliance with nuclear and environmental safety requirements, to ensure cost control, and to make the best possible use of the available resources.

Cogéma, which had the human and technical resources necessary to implement the programme, was named the prime contractor. Cogéma carries out the engineering studies, performs or subcontracts the work and operates the necessary facilities while assuming the responsibility of being the nuclear operating organization for the site.

The decommissioning operations cover three major programmes:

(1) Final shutdown which consists of rinsing and decontaminating the UP1 plant and its supporting facilities,
(2) Retrieval and repackaging of waste currently in on-site storage,
(3) Dismantling and surveillance of the facilities.

I–4.4. Organization and financing of CODEM

Figure I–8 shows the industrial organization of the project and its financing structure. CODEM is funded by its three members, 45% by the CEA, 45% by EDF and 10% by Cogéma.

As has already been stated, CODEM is the decision making, financial and supervisory entity for the decommissioning operations. In this, CODEM is supported by the customer assistance structure and carries out specific research and development work on its own behalf.

Implementation of the programme is assigned by CODEM to Cogéma as the prime contractor and the industrial and nuclear operating organization. Cogéma is therefore responsible to the safety authority. In its prime contracting role, Cogéma uses a number of subcontractors. Cogéma relies on ANDRA and

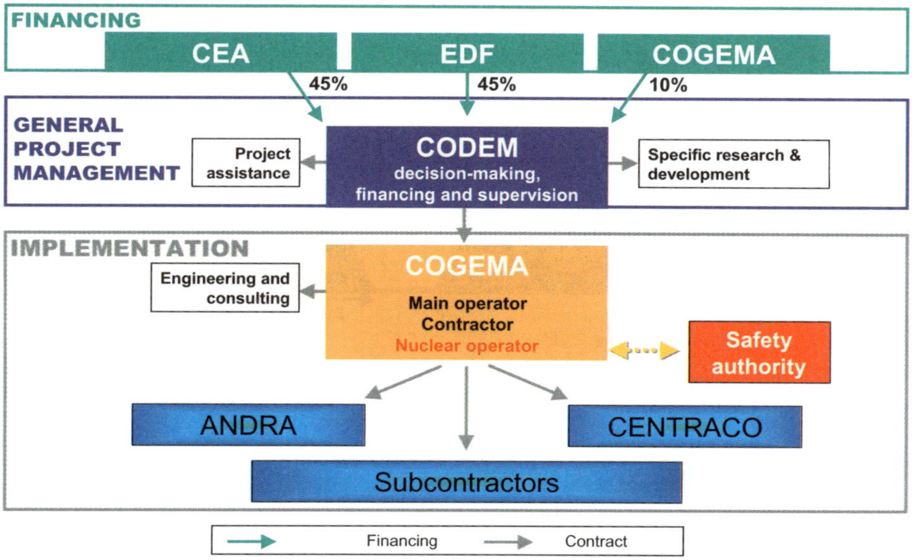

FIG. I–8. Organization and financing of CODEM.

SOCODEI for waste removal and disposal, and engages subcontractors for specific tasks.

The relationship between the client, CODEM and the service provider, Cogéma, was defined in an agreement signed on 12 February 1998. The work was subdivided contractually into lots between CODEM and Cogéma under the terms of ten operating contracts, fifteen equipment and works contracts, a waste storage contract, and a design and management contract. The financial responsibilities for each contract are specified based on UP1 production data. Accordingly, each of the members of the joint venture is assigned financial liability on the basis of the services received and the waste materials produced during the plant's operational phase. Cogéma in turn awards design and task management contracts to a group of firms including SGN, EDF and Technicatome. Subcontracts for equipment and service providers are proposed for tender by Cogéma under CODEM supervision.

CODEM's managerial role is based on a comprehensive operating scenario that defines the programme's schedule and overall cost. The scenario is revised annually and is based on a general waste survey with the operational reference consisting of the fundamental documents with which CODEM and Cogéma must ensure compliance.

As shown in Fig. I–9, work began on 1 January 1998. The final shutdown programme predominates in the early years, from 1998 to 2002, after which it is

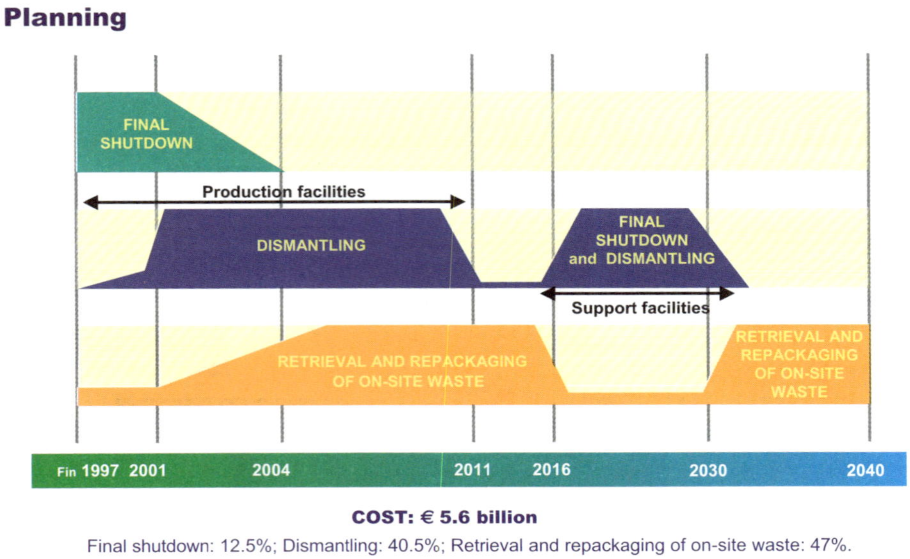

FIG. I–9. The CODEM operating scenario.

superseded by the surveillance and dismantling programme of the production facilities from 2001 to 2011. The final shutdown and dismantling programmes for the supporting facilities is implemented beginning in 2016.

The waste retrieval and repackaging programme also began with a few limited operations in January 1998 and will become increasingly important over the next decade. Its scope will shrink after 2016 until major work resumes in 2030 and is scheduled for completion by 2040. The total cost of the decommissioning project is € 5.6 billion, including 12.5% for the final shutdown operations, 40.5% for dismantling and 47% for waste retrieval and repackaging, including a 15% project margin.

I–4.5. Description of the programme

The scope of the CODEM operations at Marcoule covers the UP1 reprocessing plant, the vitrification facility, and the liquid and solid waste treatment units. The decommissioning operations are subdivided into three main programmes.

I–4.5.1. Final shutdown of the UP1 plant and related facilities

The final shutdown operations include decontamination, firstly through rinsing with various reagents, then through hands-on decontamination by the operating organizations. During this phase, the UP1 reprocessing plant is in a state very similar to its operating configuration. The main objectives of the equipment decontamination phase are to:

(a) Place the facilities in a safe condition;
(b) Reduce the residual activity level in order to limit the committed occupational dose during the dismantling phase;
(c) Reduce the need for heavy remote handling operations;
(d) Diminish the source term;
(e) Optimize subsequent waste management;
(f) Reduce surveillance requirements.

I–4.5.2. Retrieval and reconditioning of on-site waste

This programme addresses the issues of recovery, characterization, sorting, processing if necessary, repackaging and providing interim on-site storage or transfer to a disposal site of a wide variety of waste generated during 40 years of operation of the UP1 plant. The principal waste forms include spent fuel structural waste, vitrified fission product canisters, bituminized waste drums from the liquid waste treatment station, water treatment process waste, and scrapped equipment produced during plant operation and maintenance. The objectives of the programme are to transfer the waste to the ANDRA disposal site as soon as possible in order to minimize interim storage requirements and to provide safe interim storage of the waste and residue pending removal.

I–4.5.3. Dismantling and surveillance of plant facilities

This programme covers all the actions intended to reduce the surveillance requirements by limiting the residual activity containment systems as much as possible. It involves not only the reprocessing plant but also the supporting facilities (liquid waste treatment, solid waste conditioning and interim storage), as well as the equipment and facilities built for CODEM operations. The objectives of the programme are to decommission the facilities by dismantling processing equipment and treating residual hot spots to the level of an 'installation on the environmental protection list', eliminating radiologically restricted access zones and setting up suitable surveillance provisions.

I–4.6. Final shutdown of the UP1 reprocessing plant

This phase concerns the spent fuel reception, interim storage and decladding facilities (cooling pools, mechanical disassembly cells), as well as the chemical processing facilities in UP1 designed to separate the actinides and fission products and recover the uranium and plutonium.

Over 500 rooms and cells are involved, including 120 gloveboxes, with 1250 t of equipment to be removed from the building, 550 m^3 of tanks to rinse and 3500 m^3 of pools to be drained. The total floor area of the buildings is 18 000 m^2.

In anticipation of final shutdown operations at the beginning of 1998, feasibility studies were undertaken in 1992. Equipment data were compiled from mid-1993 to mid-1994 for preliminary design work in 1994 and 1995. During the same period, research and development work was carried out on decontamination products. Selected reagents were tested at the end of 1998 in a representative tank at the Marcoule pilot reprocessing facility. The development phase lasted from 1996 to mid-1997 and the preparatory work carried out during the second half of 1997 allowed the actual decontamination operations to begin in January 1998 for a five year period.

I–4.6.1. Liquid and solid waste from the final shutdown operations

The waste generated during the final shutdown of UP1 is processed and conditioned in the existing facilities, liquid waste in the liquid waste treatment station and the Marcoule vitrification facility, and solid waste in the equipment decontamination unit and the solid waste conditioning facility. These units will continue to operate at normal rates throughout the final shutdown phase. The waste produced during the final shutdown decontamination phase is processed and conditioned separately for each waste category:

(a) Waste suitable for surface disposal is conditioned on-line and transferred to the ANDRA disposal site in northeastern France.
(b) Waste containing appreciable amounts of plutonium is treated in the same way as during the operating period of the plant. The most highly radioactive waste, which represents only a small volume, is placed in interim storage with the existing waste awaiting treatment under the retrieval and repackaging programme.
(c) Some waste with very low contamination levels is suitable for incineration or melting.

I–4.6.2. Processes and techniques

I–4.6.2.1. Rinsing

The processing equipment is rinsed with a variety of reagents. These include conventional reagents such as nitric acid and sodium hydroxide, as well as the specific reagents potassium permanganate in sodium hydroxide, tartaric acid, hydrofluoric acid or cerium (IV) in nitric acid. Rinsing is repeated until the residual plutonium content is less than 1 g and 37 GBq/m^3 with the most effective reagent.

All the liquid waste streams are neutralized, acidified and monitored. Tartaric acid is decomposed by boiling in the presence of manganese; potassium permanganate and cerium (IV) are neutralized with hydrogen peroxide; the fluorine ions are complexed with aluminium.

Most of the resulting liquid waste is concentrated by evaporation (with concentration factors ranging from 10 to 20) and transferred to the vitrification facility where it is blended with the high level fission product solutions produced by reprocessing. It is then calcined and vitrified. Blending minimizes the number of glass canisters required while ensuring compliance with the specified chemical element concentrations and maintaining the nuclear glass containment quality. The remaining liquid waste is processed in the liquid waste treatment station by evaporation and co-precipitation, and the resulting sludge is encapsulated in bitumen.

I–4.6.2.2. Solid waste treatment

All the solid waste is initially placed in standard waste packages:

(a) 118 L steel drums, concrete lined 223 L drums or 2800 L stainless steel bins;
(b) Stainless steel transport containers for waste to be decontaminated and/or cut up in the equipment decontamination facility before being placed in 2800 L bins.

Type A low level waste destined for the ANDRA surface disposal site is grouted with cement inside 223 L drums or in 4 or 5 m^3 steel or fibre cement containers. Type B waste intended for the recently completed on-site multi-purpose interim storage facility is conditioned in 218 L stainless steel drums placed inside 380 L stainless steel overpacks with intermediate radiological shielding. Type C vitrified waste is solidified in 150 L stainless steel canisters and placed in interim storage at the vitrification facility.

The activity of each package is measured by counting based on typical radionuclide spectra. The spectral characteristics are subject to change as the decontamination operations progress and are systematically reassessed by sampling and destructive measurements.

I–4.6.2.3. Nuclear measurements

The effectiveness of the rinsing and mechanical decontamination operations is assessed by conventional non-destructive methods such as dose rate measurements by gamma detectors or by passive neutron counting. More sophisticated methods are also used:

(a) Active neutron scanning to determine the quantity of residual plutonium in processing equipment such as the dissolvers,
(b) Gamma cameras to identify the most highly active zones,
(c) In situ object characterization system (ISOCS) measurements to determine the quantity of residual plutonium in the equipment on the plutonium line and the gamma emission spectra.

I–4.6.2.4. Remote operation and robotics

Remote operating techniques are implemented using commercially available equipment qualified by the CEA and specialized firms, including Brokk carriers and arms, SAMM dexterous arms, and telescoping masts. Five remotely operated devices are used:

(a) Three ground units with Brokk carriers, including one with a SAMM arm;
(b) An underwater unit to clean spent fuel pools consisting of a carriage, a telescoping mast and a SAMM arm;
(c) An aerial unit with a SAMM arm on a telescoping mast suspended from a rail-mounted carriage.

I–4.7. Conclusions

The UP1 decommissioning programme has now been going on for more than four years and is progressing well. Priority is given to ensuring optimum safety conditions and minimizing personnel dose exposure. These objectives are achieved by the designation of a single, responsible operating organization with long standing safety experience.

The organization set up for this programme is designed to separate the interests of the client, CODEM, from those of the service providers in order to ensure cost effective and technically viable solutions. The organization is funded under a multi-year financing plan.

The technical tasks have been fully identified, and generally correspond to known operations that have been carried out in the past. The necessary technology is generally available, including chemical engineering, mechanical equipment and robotics, and has already been implemented in a nuclear environment.

The final disposal routes for the solid waste removed from the site will be fully defined as soon as the relevant statutory requirements are published. These arrangements will cover very low level waste as well as medium and high level waste under the terms of the 1991 radwaste management law.

Arrangements have been put in place to benefit from the experience gained on this project, which is the largest of its kind carried out to date in France. The data obtained from this programme will provide a sound basis for the economic and technical assessments required for future decommissioning programmes.

I–5. THE TRANSITION PERIOD IN GERMAN NUCLEAR REACTORS

I–5.1. Introduction: Overview of installations

The following summarizes the status of German nuclear installations at the end of the year 2001:

(a) Nineteen commercial NPPs were in operation and contributed up to a third of overall electricity production in Germany. In addition, some research reactors and a few fuel cycle installations were in operation.
(b) Sixteen NPPs and prototype reactors, 11 larger research reactors, 18 smaller research reactors and critical assemblies as well as 6 fuel cycle facilities were permanently shut down.
(c) Two of the 16 power reactors, 18 research reactors and critical assemblies and one of the fuel cycle facilities have already been decommissioned. The sites of these two power reactors (KKN Niederaichbach and HDR Grosswelzheim) were cleared and restored to green field condition. Two of the power reactors and three research reactors are in SE. At the other installations dismantling is in progress with the aim of reaching green field conditions in most cases.

I–5.2. Licensing requirements

The German Atomic Energy Act (Atomgesetz, AtG) is the basis for regulation of any nuclear activity. Requirements for radiation protection measures are laid down in the Radiation Protection Ordinance (Strahlenschutzverordnung, StrlSchV) and licensing procedures are defined in the Nuclear Licensing Procedure Ordinance (Atomrechtliche Verfahrensverordnung, AtVfV).

According to the Atomic Energy Act, the competent regulatory authority for licensing and supervision lies with the federal state (Land) in which the nuclear installation is located. The Federal Ministry for the Environment, Nature Conservation and Nuclear Safety (Bundesministerium für Umwelt, Naturschutz und Reaktorsicherheit, BMU) has a supervisory function and may give directives to the state authorities. The subordinate ordinances to the Atomic Energy Act, especially the StrlSchV and the AtVfV, apply to the construction and operation of NPPs, as well as to decommissioning. Because the regulatory framework, as well as guides and safety standards, are mainly oriented towards construction and operation they have to be applied in the same way for the transition and decommissioning phases. In order to give guidance on the currently applicable regulatory framework regarding decommissioning the Guide to the Decommissioning of Facilities as defined in Section 7 of the Atomic Energy Act was issued in June 1996 and is currently being revised. The submission of a preliminary decommissioning plan (feasibility study, estimation of waste, etc.) is a precondition for granting an operating licence. The decommissioning plan must be revised every five years.

Section 7 of the Atomic Energy Act is the basis for construction, operation and decommissioning of nuclear installations. A separate licence is required for each step in a plant's life: construction, operation, decommissioning, SE and dismantling. Section 7 para. 3 of the AtG states: "The decommissioning of an installation as defined in para. 1 as well as the SE of a finally decommissioned installation or parts thereof shall require a licence". Among many other things, a licence defines the scope of the operations and procedures it permits. The prerequisites for obtaining a decommissioning licence are essentially the same as for construction and operation of NPPs and have to be applied as appropriate:

(a) Reliability and professional qualification of responsible personnel;
(b) Required qualification and knowledge of all other personnel;
(c) Precautions against damage resulting from decommissioning activities, reflecting the current status of science and technology;
(d) Financial provisions (insurance) for liability;
(e) Physical protection from third party actions;
(f) Environmental considerations.

It should be noted that public involvement is part of the licensing procedure for awarding decommissioning licences but is not required for amendment of existing operating licences.

I–5.3. Decommissioning strategy

Under the Atomic Energy Act the choice of decommissioning strategy and when to apply for a decommissioning licence is left to the licensee. Generally, the strategies of immediate or deferred dismantling are both considered viable in Germany and both have been chosen in the past. Decisions for deferred dismantling were taken mainly for financial reasons. Nevertheless, some decommissioning activities are necessary immediately after reactor shutdown to ensure appropriate conditions for SE. The majority of the licensees decided in favour of immediate dismantling of the entire plant including the setting up of plans for future restoration of the site.

Usually the entire decommissioning process will be divided into several project phases with application for and granting of individual licences. It has been proven advantageous for both the licensee and the regulatory body to initially apply for a licence covering deregulation of operational procedures, in particular a reduction of the requirements for control room staff, suspension of regular testing and inspection of operational safety equipment, etc.

The goal of decommissioning is to clear (release) the site with no restrictions because of radiological hazards, with the remaining buildings either being used for other purposes or completely restored to green field status, allowing the site to be used for other purposes as defined in the land development plan. Decommissioning can also be considered complete if the remaining plant structures are used as a nuclear facility under a new licence.

I–5.4. The post-operational phase (transition period)

The transition period from final shutdown to the start of dismantling (decommissioning) is termed the 'post-operational phase' in Germany. This indicates that this phase in a plant's life is still governed by the operating licence. In the post-operational phase, which may last up to around 3 years for a larger NPP, the following work may be carried out:

(a) The fuel elements are unloaded and transported from the reactor to an on-site or a centralized storage facility. In Germany, centralized fuel element storage facilities are available at Gorleben and Ahaus. However, an agreement has been reached between the German power utilities and the Federal Government that fuel element storage facilities will be constructed at the sites of all NPPs. The licensing procedures for this are in progress. It should be noted that for high temperature reactors with a pebble bed core (AVR and THTR) the unloading of all the fuel elements is not a requirement of the operating licence. Unloading of the cores of those two reactors was only possible after the first decommissioning licence had been granted.
(b) The operating wastes are removed from the plants and sent for conditioning.
(c) The circuits are drained. In most cases this is followed by decontamination of the circuits in order to reduce the dose rates for subsequent dismantling work (as part of the decommissioning licence).
(d) Systems and equipment which the regulatory authorities have agreed are no longer required are taken out of operation. This helps to reduce costs and staffing requirements.
(e) The decommissioning strategy is defined and the first stages of decommissioning are planned in detail. All documents needed to apply for the first decommissioning licence are prepared. The operating organizations

usually plan the licensing procedure in such a way that granting of the decommissioning licence coincides with the end of the post-operational phase. However, this cannot always be achieved in practice.

(f) Dismantling work can only commence after the decommissioning licence has been granted. However, some German federal states have issued amendments to existing operating licences in order to allow partial dismantling of specific systems before the comprehensive decommissioning licence is issued.

It should be noted that the plant owner retains responsibility during the entire post-operational and decommissioning phases, until the plant has been dismantled and the site has been cleared. This means that the personnel who operated the plant usually plan the decommissioning phase.

I–5.5. Case studies

This section deals with the post-operational phase of various plants. A great deal of additional information is available on the subsequent decommissioning phases but is omitted here because it is outside the scope of this report.

I–5.5.1. Würgassen NPP (KWW)

The NPP at Würgassen belongs to E.ON Kernkraft GmbH. It operated until 1995 and received its first decommissioning licence in April 1997. The post-operational phase lasted about two years. During this period the operating licence remained in force and the following work was carried out:

(a) General preparations for decommissioning began.

(b) All irradiated fuel elements were removed from the site and transported to a central storage facility (Fig. I–10). This reduced the radioactive inventory of the plant by over 99%.

(c) The cooling towers were demolished (Fig. I–11). This was possible because they did not belong to the nuclear part of the facility, i.e. they were not included in the plant's nuclear licence. Therefore it was not necessary to obtain a licence according to Section 7 para. 3 of the Atomic Energy Act for this work. The rubble (21 500 t) was recycled on the site or in road construction, the 600 t of reinforced steel was recycled by melting. Because the cooling towers had not been part of the nuclear facility it was not necessary to clear this material.

FIG. I–10. Removal of irradiated fuel elements from the core of the Würgassen NPP.

FIG. I–11. Demolition of the cooling towers of the Würgassen NPP.

Planning carried out during the post-operational phase revealed that direct dismantling would be more desirable than SE. This also helped to maintain employment levels and the NPP continued its role as a major employer in a region which lacks a strong infrastructure. As shown in Fig. I–12, during the post-operational phase the number of plant staff gradually dropped from around 330 (during operation) to below 200 at the beginning of decommissioning. During decommissioning it has increased to between 450 and 500, about 130 E.ON Kernkraft GmbH employees and the rest contractors [I–1].

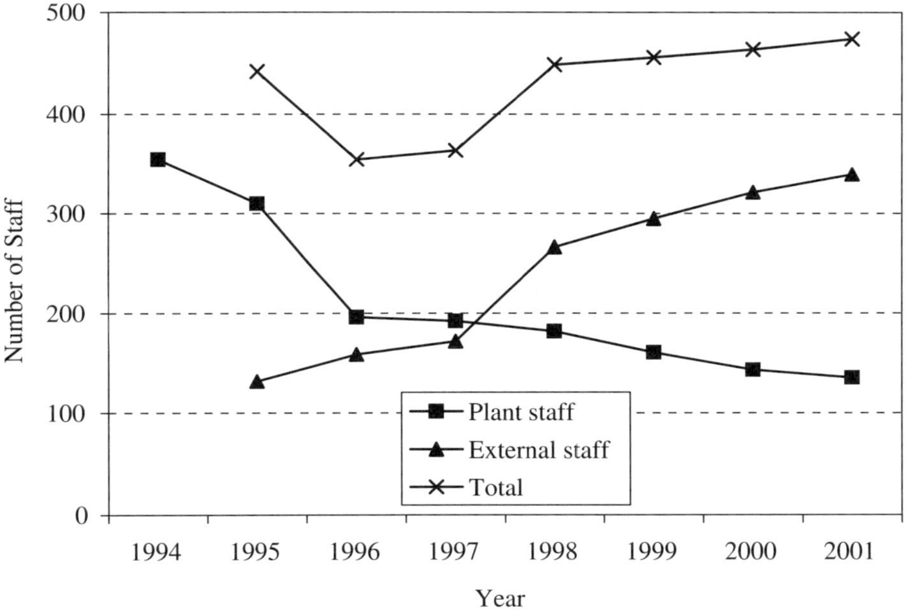

Year	Plant staff	External staff	Total
1994	355		
1995	310	132	442
1996	196	159	355
1997	192	172	364
1998	182	267	449
1999	161	295	456
2000	143	321	464
2001	135	339	474

FIG. I–12. Development of KWW staff numbers — plant staff (excluding trainees and temporary workers) and external staff.

A significant amount of information was made available to the public in the area around the Würgassen NPP during the post-operational phase. An exhibition about the decommissioning of NPPs was developed and placed in the visitor centre on the site. New information leaflets and CD-ROMs were used to keep all those interested informed of the progress of the decommissioning.

I–5.5.2. Lingen NPP

The NPP at Lingen was a 240 MW(e) BWR with a fossil fuel fired super heater. Construction commenced in late 1964 and the plant operated between 1968 and 1977. The plant was shut down in January 1977 because of technical problems with the steam to steam heat exchangers. Technical improvements to the plant as well as its conversion to a conventional power plant were investigated but not considered to be feasible.

In June 1983 a licence for the establishment and operation of an SE was applied for. This was granted in November 1985. The Lingen NPP has been in an SE state since 1988 with a planned duration of 25 years.

The post-operational period lasted from final shutdown in 1977 to the issuing of the first decommissioning licence in 1985. During this time, the following work was carried out:

(a) Unloading of the fuel elements was approved under the operating licence and lasted from 1981 to 1983.
(b) Various installations were reduced in size or dismantled, including evaporator plant and control gas systems.
(c) Highly radioactive parts such as control rods, fuel channels etc. that were activated during the operation of the reactor were placed in the reactor pressure vessel for storage during the SE. Other components with less activity were stored in the fuel storage pool or the steam separator pool.
(d) As it was planned to maintain the SE without low pressure in the buildings, leakage from the entire building complex designated for the SE was measured. The reactor containment was proven to be airtight.
(e) The contamination was characterized and ^{60}Co was identified as the dominant nuclide. In the activated biological shield the most predominant nuclides are ^{152}Eu, ^{60}Co, ^{154}Eu and ^{134}Cs.
(f) After the removal of all stored components the fuel storage pond was drained, cleaned and decontaminated in 1984. The primary circuit was drained and dried between 1984 and 1987.
(g) The ventilation system that had been used during the operational phase was switched off (which also ended the ventilation of the primary loop).

A new ventilation system for the SE period was installed and put into operation in 1987. Redundant openings in the buildings were sealed.

The plant employed approximately 160 people during operation. This increased to over 200 during preparation of the SE (1985–1987) and then decreased rapidly to a few people in 1988. Any necessary work in the plant is carried out by contractors.

I–5.5.3. Greifswald NPP

The NPP at Lubmin near the city of Greifswald consisted of five operational Soviet type WWER PWRs (one in trial operation) plus 3 reactors which were under construction when the decision for final shutdown was taken in 1990. The post-operational phase lasted from 1990 to 1995. The first decommissioning licence was issued in June 1995.

During the post-operational phase one of the major tasks was to develop a decommissioning plan because the final shutdown of the plant was unplanned and there was little experience in the decommissioning of WWER type reactors. This was complicated by the difficult transition from regulation under the laws of the former German Democratic Republic to the Atomic Energy Act of the Federal Republic of Germany. In developing the decommissioning concept, the following points had to be taken into consideration [I–2]:

(a) Planning the application for a decommissioning licence;
(b) Maintaining the necessary functions of the remaining operations to achieve the planned objectives, i.e.
    — Ensuring that the fuel is not in a critical condition,
    — Removing the radioactive decay heat from the fuel,
    — Maintaining radiation protection standards,
    — Safeguarding all residual radioactive materials;
(c) Providing storage capacity for the fuel and radioactive wastes removed from the plant;
(d) Reducing the operating expenditures for the plant;
(e) Minimizing radiation exposure to the staff at the plant and to the environment during the dismantling;
(f) Considering plant specific features influencing the dismantling;
(g) Minimizing radioactive wastes during dismantling and safe storage of the wastes.

This resulted in the following tasks having to be carried out during the post-operational phase:

(1) Monitoring of the nuclear facilities using a three shift system;
(2) Environmental surveillance;
(3) Security of the NPP;
(4) Handling, transportation and storage of fuel elements;
(5) Conditioning of radioactive operational (and decommissioning) wastes;
(6) Preparation and use of tools and techniques, including remotely operated equipment, for the dismantling and disassembly of plant components;
(7) Application of decontamination techniques;
(8) Development of procedures for clearance of material;
(9) Establishment of a mass and activity inventory.

Construction of a storage facility (Zwischenlager Nord) on the site of the Greifswald NPP enabled most of these tasks to be completed and the dismantling of the plant to be separated from the further segmentation and waste conditioning. This is being done on-site and hence does not impair the dismantling. It also enables the dry storage of spent fuel.

I–5.5.4. Gundremmingen NPP

Unit A of the Gundremmingen NPP was built between 1962 and 1966 as the first of three demonstrator NPPs in Germany. It had a BWR rated at 250 MW(e) and operated between 1966 and 1977. In January 1977 it was taken out of operation after an incident. The decision to decommission it was taken in 1980 because of the high costs for required repair and upgrading and because the construction of units B and C with 1300 MW(e) was progressing well.

The post-operational phase lasted from 1980 to 1983 when the decommissioning licence was granted and decommissioning work began with dismantling in the turbine hall. The post-operational phase was also used for detailed planning of the decommissioning (as well as for removal of spent fuel, operational wastes, etc.). This proved that decommissioning could be done:

(a) Without a long period of time in SE,
(b) On a reasonable economic basis,
(c) With low doses to the personnel,
(d) With negligible release of radioactivity to the environment,
(e) With little secondary waste.

During the post-operational phase (as well as during decommissioning) it was advantageous to have the infrastructure and the logistics available from the two other operating reactors at the site.

I–5.5.5. Karlsruhe FR2 reactor

The Karlsruhe FR2 reactor was a research reactor with 44 MW(th) located at the Karlsruhe research centre. Its final shutdown took place in December 1981. It was decided to put the plant into SE after partial dismantling.

During the post-operational phase some work towards dismantling and SE was permitted under amendments to the existing operating licence. This work included:

(a) Removal of fuel elements from the core,
(b) Removal of the primary coolant D_2O for further use at other nuclear facilities,
(c) Shutdown of a number of auxiliary and control systems,
(d) Draining of other circuits and systems,
(e) Sealing of openings,
(f) Simplification of systems that had to be kept operational.

Further work was only permitted after granting of the first decommissioning licence on the basis of Section 7 para. 3 AtG. A sequence of six decommissioning licences led to the current state where only the reactor itself is in a condition of SE, some buildings have been dismantled and the reactor building is accessible. Final dismantling is planned for no earlier than 2015. The first plans need to be presented in 2010.

I–5.5.6. Special cases

(a) Karlsruhe research reactor:
The research reactor (Mehrzweckforschungsreaktor (MZFR)) at the Karlsruhe research centre was a 200 MW(th) (50 MW(e)) pressurized heavy water reactor which was operated between 1966 and 1984. It was not possible to drain and dry the heavy water systems (outside and inside the reactor building) before the first of a series of decommissioning licences had been granted. In this case, the licences will be issued as a series of eight sublicences for each step of the process until green field conditions are achieved. [I–3].
(b) Jülich reactor:
This 15 MW(e) reactor, the Arbeitsgemeinschaft Versuchsreaktor, Jülich (AVR) was a high temperature helium cooled reactor with spherical fuel elements (pebble bed). This experimental reactor operated between 1967 and 1988. After final shutdown it was decided to place the plant in SE.

The licence for decommissioning, unloading of the reactor core, dismantling of components and SE was granted in March 1994.
Complete defuelling had not been an operational procedure. Therefore it was not possible to remove the approximately 130 000 fuel elements under the operating licence and defuelling had to be part of the first decommissioning licence. (This also applied to the THTR 300 at Hamm-Uentrop which was a 308 MW(e) pebble bed high temperature reactor.)
Because of the experimental character of the AVR the importance of the personnel's good knowledge of the plant and its operating history was considered very important in the planning for decommissioning. Therefore it was regarded as better to dismantle the plant without an SE period. A new licence will be required for dismantling.

I–5.5.7. Conclusions

The cases presented in this section demonstrate that the scope of work during the transition period or the post-operational phase in Germany is determined by the content of the operating licence. As this transition period is covered by the operating licence it is not possible to do any work which is specific to decommissioning. Examples of work which can be done include:

(a) Planning for decommissioning (usually in great detail for the first stages of decommissioning and in less detail for later steps),
(b) Unloading of fuel elements,
(c) Removal and conditioning of waste,
(d) Minor dismantling work,
(e) Rationalization of systems.

I–6. THE TRANSITION PERIOD IN ITALY

I–6.1. Introduction

The transition phase between plant operation and decommissioning in Italy can be divided into two periods:

(1) Following a moratorium on the use of nuclear power that started with a referendum following the Chernobyl accident.
(2) The time period since July 1990 when the Government decided to close the remaining operating NPPs, the 270 MW(e) PWR at Trino and the

860 MW(e) BWR at Caorso. The other two Italian plants, Latina and Garigliano, had already been closed.

All the NPPs were owned and operated by the state owned ENEL Company, the National Electricity Board. During the moratorium the NPPs were put into a 'cold shutdown' condition and long term preservation of systems was carried out. To save costs, only reversible actions and checks on 'out of service' systems were carried out. All these actions were carried out under the current operating licences with formal documents and procedures from the operating handbook as approved by the Control Authority.

The plants retained the same organization as during operation, with almost the same number of staff. Staff competencies were preserved in readiness for a restart. However, most of the people were working on modifications required to put the plants into a state of long term preservation.

A period of major reorganization followed the Government's decision to close the plants. This included reductions in the number of staff employed, changes in the decommissioning strategy, reorganization of the company's structure and implementation of changes to the laws pertaining to nuclear energy.

The previous nuclear regulatory law (DPR[3] 185 of 1964) did not provide a clear legislative framework for decommissioning, making it difficult to plan for. Such a framework was put in place in 1996 when a new law (230/95) was issued. This law embodied a number of recent Euratom directives concerning radiation protection of workers and the population. The law was implemented in 2000 (241/00).

This change in the law created delays in decommissioning activities due to the need to review documents required by the authorities for issuance of decommissioning licences. A change of decommissioning strategy from safestore to 'immediate' dismantling, i.e. dismantling in a single stage as opposed to deferred dismantling after a number of years of safestore, caused further delays. This new decommissioning strategy was set down in Ministry of Industry guidelines of December 1999 and was put into law by a Ministry of Industry decree in May 2001. The decree stated that:

(a) The four Italian NPPs would be decommissioned by 2020.
(b) A national repository for irradiated fuel and radioactive wastes would be operational at the beginning of 2009 (the site would be chosen by 2005).

[3] DPR: Decree of the President of the Republic.

(c) Wastes arising from past operation of the NPPs would be treated/conditioned by 2010.
(d) Spent fuel would not be reprocessed. Fuel assemblies would be stored in dry dual purpose casks and temporarily stored at the NPP sites awaiting transfer to the national repository. Only a small amount of fuel would be sent for reprocessing under existing contracts.

In March 1999 a new legal requirement for competition in the Italian electricity market was imposed. At the same time ENEL was restructured and transformed into a holding company with a number of independent subsidiary companies. At the end of 1999 all aspects of nuclear energy (production, research, etc.) were brought together in a company called SOGIN (Nuclear Facilities Management Company). The mission of this new company is to:

(1) Manage the post-operation activities of the four Italian NPPs;
(2) Plan and manage the decommissioning of the NPPs in accordance with the new strategy;
(3) Manage the end of spent fuel reprocessing;
(4) Manage the restoration and reutilization of the NPP sites;
(5) Develop business activities in national and international markets (nuclear services abroad, environmental studies, etc.) by using past operational and newly acquired decommissioning experience.

In November 2000 responsibility for SOGIN was transferred to the national treasury, while ENEL followed a path to privatization. SOGIN is a joint stock company but all the stock belongs to the treasury. SOGIN employs about 600 people, 200 at its headquarters and 400 at the NPPs.

To enable SOGIN to carry out its activities, funds which had been reserved for dismantling activities at NPPs were transferred to it from ENEL. SOGIN also benefits from a government contribution taken from electricity charges to all customers.

I–6.2. Redeployment of redundant personnel

As a state owned organization, ENEL made every effort to employ personnel displaced from the NPPs in other areas of its business, e.g. in conventional plants, research laboratories, training schools and the distribution organization:

(a) Transfer requests to neighbouring conventional plants were approved. Transfer was easier for operations and maintenance department personnel who have transferable skills. For example, three thermoelectric plants were in operation near Caorso, and a new conventional plant had been constructed near Trino. Many operations department staff were employed at these conventional plants.

(b) An early retirement scheme was also provided for staff.

(c) Some staff were given new and different assignments either at the NPP or elsewhere, but within the ENEL organization. These new functions differed from their normal duties but utilized their existing skills.

This situation entailed a more flexible concept of the workplace. Individuals were not constrained by their former roles but allowed more flexibility to apply their knowledge, experience and skills.

These efforts were discussed and developed in conjunction with the trade unions. Therefore ENEL retained a greater number of employees in the plants than strictly required for routine NPP activities.

Staff morale is very important during the transition phase. The uncertainty resulting from a decision to permanently shut down or from the decommissioning strategy after closure is extremely frustrating for plant staff and can threaten morale. This can lead to a loss of qualified people and 'historical memory'. Personnel remaining at the plant have to change their mentality when they are faced with the reality that they are dismantling what they had constructed and managed until a few days before.

Experience has shown that there were fewer problems motivating the chemical and health physics and maintenance departments than the operations department. In fact, chemical and health physics department personnel could be involved in many 'new' activities such as plant radiological characterization campaigns, waste treatment, conditioning, and radiological characterization. The decrease in regular maintenance activities allowed maintenance department personnel to be assigned to dismantling or system retirement and reconfiguration activities.

Greater difficulty was found in motivating operations department personnel. The reassignment of shift workers to ordinary working groups, for example, is difficult, inasmuch as they know that shifts will be reduced or eliminated in the near future. Only a few highly specialized people such as shift supervisors or some reactor desk operators can be involved in the 'new' pre-decommissioning activities.

The creation of mixed working groups or teams in which experienced staff with knowledge of the NPP work together with new staff who have less difficulty with the new objective of dismantling the plant can be a good approach to solving the motivational issues. For example shift supervisors and/or maintenance staff can be moved from their original department to new ones which include new resources. Together they can then plan the future pre-decommissioning activities.

I–6.3. Experience of the Caorso and Trino NPPs

The Caorso NPP, in northern Italy on the River Po about 80 km southeast of Milan, is an 882 MW(e) General Electric BWR reactor. It started commercial service in December 1981 and was shut down at the end of its fourth fuelling cycle on 25 October 1985 after a total electricity production of about 29 000 GW·h. All the fuel (1032 fuel bundles) is stored in the plant's fuel pool. The plant's staff currently numbers about 170 people.

The Trino NPP, also in northern Italy by the Po River, is a 270 MW(e) Westinghouse PWR. It entered service in January 1965 and was shut down at the end of its ninth fuel cycle on 21 March 1987 after a total electricity production of about 25 000 GW·h. There are still 47 irradiated fuel elements in the pool. The Trino NPP has a staff of about 90 people.

Plant personnel were engaged in the design and planning of many of the activities which had to be performed following the decision to close the NPP. Many systems were no longer necessary after shutdown, particularly after defuelling, including uncontaminated systems and systems not coming under the technical requirements of the operating licence. Such systems were retired after drainage of the fluids contained in them (water, oil, etc.), their electric segregation and the removal of hazardous materials such as asbestos (see Section I–6.3.3). The aims of this were to:

(a) Reduce the risks from hazardous materials;
(b) Reduce operating and maintenance costs (e.g. power and fuel consumption, surveillance, maintenance, etc.);
(c) Allow requalification and retraining of personnel to new duties.

Other actions were taken on systems that were still needed for future decommissioning activities after shutdown but were 'over-specified' for their future functions and too costly to operate and maintain in their old configurations (see Section I–6.3.2).

Other actions which can be taken ahead of the main decommissioning activities could be the decontamination of some systems, mainly the primary system (with the aim of dose reduction during routine inspections), dismantling of new components and asbestos removal. Any resalable/valuable materials can be disposed of (see Section I–6.3.3). It should be pointed out that these activities come under the scope of the operating licence and are mainly done as 'plant modifications'. However, they have to conform to the operating procedures and always be carried out under the supervision of the control authority as a decommissioning licence is not in force.

I–6.3.1. Reduction of conventional and radiological hazards

The number of conventional and radiological hazards has been reduced through the following:

(a) Identification of the nature and quantity of conventional and radiological hazards present in the plant;
(b) Classification of significant hazards;
(c) Identification of those actions necessary to eliminate or reduce hazards;
(d) Identification of those plant systems still required to maintain safety;
(e) Estimation of the time needed and the cost of modifications, either in absolute terms or through costs–benefit analysis;
(f) Implementation of the modifications.

The activities described below originated from the above mentioned analyses.

I–6.3.1.1. Reduction of fire hazard and risk of uncontrolled release of toxic substances

(a) Removal of all the turbine lubricating oils together with the storage tanks;
(b) Removal of the main transformers and the oil contained in them;
(c) Removal of all the transformers containing PCB;
(d) Removal of the antifreeze, lubricating oil and fire fighting foam from the emergency diesels (Trino NPP);
(e) Reduction of diesel fuel stocks (Trino NPP);
(f) Removal of asbestos and rock wool from the conventional zone of the plant, and decontamination;
(g) Removal of asbestos and rock wool (Fig. I–13) from the controlled zone of the plant, and decontamination.

I–6.3.1.2. Reduction of the radiological risk

(a) Radiological characterization of all the plant systems;
(b) Characterization of all the drums present in the plant by γ spectrometry and other techniques with appropriate scaling factors (Fig. I–14);

FIG. I–13. Asbestos removal at the Trino NPP.

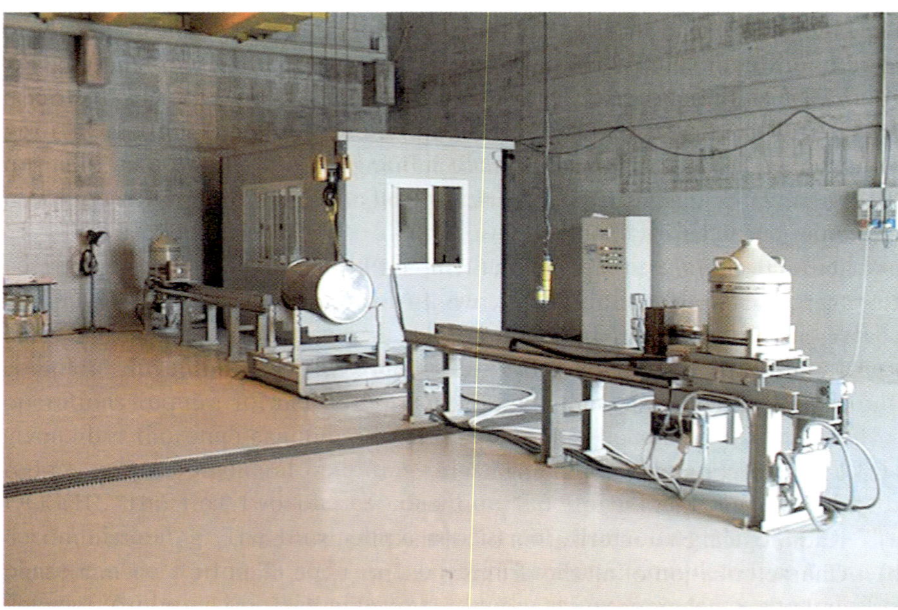

FIG. I–14. Drum radiological characterization at the Trino NPP.

(c) Rationalization of all drums in the storage building to free space and facilitate inspection (Fig. I–15);
(d) Identification and classification of all radioactive waste within the plant;
(e) Transfer, where possible, of all radioactive waste into drums;
(f) Maintenance of all the systems and plant buildings necessary for nuclear security and radiological containment to maintain the efficiency and reliability of the equipment and to avoid degradation of the components and structures.

I–6.3.2. Reduction of operating costs

All systems have been reviewed and classified into the following categories:

(a) Systems which must carry out their principal function;
(b) Systems which have a radiological containment function;
(c) Systems connected to those which have a radiological containment function;

FIG. I–15. Revised drum storage arrangements at the Trino NPP.

(d) Support systems;
(e) Systems with no future function.

The power consumption, fuel and maintenance costs of the first four groups have been analysed, taking into account the safety requirements of the plant. As a result it is evident that some costs can be reduced or avoided by making small changes or modifications. The following improvements were identified:

(1) Modifying the heating system to meet the new requirements by replacing the old boiler with two new, cheaper and more versatile ones (Trino);
(2) Modifying some systems to meet the requirements of the new plant configuration, e.g. a reduction in the consumption of water by the plant allowed changes to be made to the pump operation, saving energy (Trino);
(3) Reconfiguring the power supplies to maintain enough power for safety, as well as some electrical appliances (Trino);
(4) Reducing the running time of the containment ventilation system (Trino);
(5) Demolition of the weather station and replacing it with the Minisodar, bringing a reduction in the maintenance costs of the weather station as extraordinary maintenance would have been necessary to the 40 year old weather station (Trino);
(6) Removal of the nitrogen atmosphere from the tanks in the controlled zone, eliminating some operating and maintenance costs (Trino);
(7) Reduction in the running time of the ventilation fan in the auxiliary (non controlled zone) and turbine building (Caorso);
(8) Optimizing the service water systems and reducing their operating time (Caorso);
(9) Rationalizing the operation (light intensity and operating time) of the external and internal lighting in plant areas (Caorso);
(10) Rationalizing the temperature of the plant ventilation systems (Caorso).

Some systems which will have no function in the future have either been retained in order to maintain their structural integrity or dismantled, e.g.

(i) Demolition of some buildings with a consequent reduction in air conditioning costs;
(ii) Electrical isolation and dismantling of systems in the turbine building;
(iii) Shutting down of many systems with a consequent reduction in maintenance operations, allowing personnel to be transferred to dismantling activities.

As a result of the changes outlined above, electricity consumption at Caorso fell at the rate shown in Fig. I–16 (The trend at Trino is shown in Fig. 5 of the main text).

The above changes led to a 20% reduction in the Trino NPP's annual operating costs. The cost of the modifications was amortised over one year. The total benefit of these changes is estimated at approximately € 8 million over 10 years. The revision of Trino's emergency plan enabled a reduction in both the internal and external resources required to support it.

I–6.3.3. Removal and sale of components

The following equipment was able to be removed from the NPPs:

(a) Trino: main 3 kV diesel generators that supplied the emergency core cooling system;
(b) Trino: metal and plastic contained in the cooling towers;
(c) Trino: main transformers;
(d) Trino: other items not necessary for the safety of the plant;
(e) Caorso: main generator;
(f) Caorso: the turbine building's closed cooling water system;
(g) Caorso: main transformers and bus way.

A total of 295 unused and new fuel elements were sold. This included both the fuel that had never been put into the reactor and the fuel that had

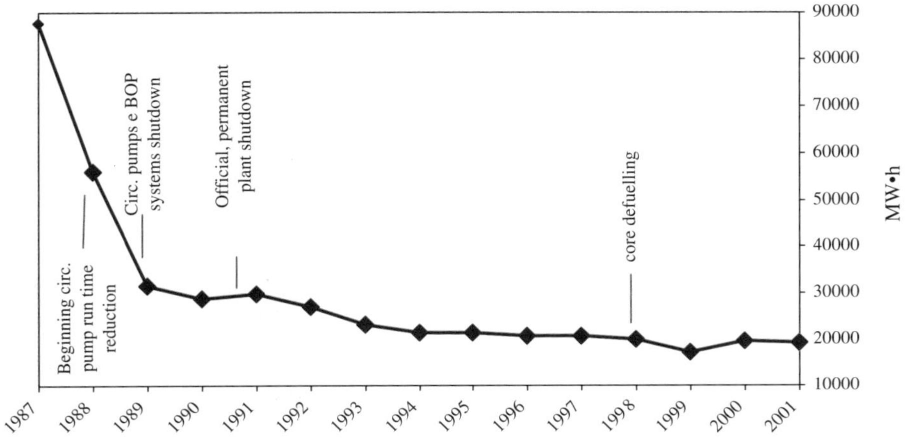

FIG. I–16. Electricity consumption trend at the Caorso NPP.

been put into the reactor but never irradiated (this was decontaminated prior to sale) (Fig. I–17).

At Trino the spare control rods were dismantled in the NPP's workshop (Fig. I–18). The Ag, In, Cd alloy was sold after dismantling. The aluminium and copper from the main transformers and associated electric bars were also sold (Figs. I–19).

FIG. I–17. Fresh fuel decontamination for shipping from the Trino NPP.

FIG. I–18. Disassembling of spare control rods at the Trino NPP to extract the Ag, In and Cd alloys.

FIG. I–19. Aluminium sold from electric bars at the Trino NPP.

I–6.3.4. Reduction and requalification of personnel

As ENEL owned the Caorso and Trino NPPs at the time of the CIPE resolution some workers were transferred to other plants at their own request. The remaining workers began discharging the fuel. Figure I–20 shows the number of employees at Trino over this period of time; Figure I–21 shows the same for the Caorso NPP.

Some workers faced the difficult reality of dismantling what they had constructed, and until recently managed. This called for a change in mentality. Moreover some people were still obliged to continue working shifts. To make the most of their abilities in preparing the necessary documentation to obtain authorization for the dismantling (and the above mentioned modifications), an early objective was to reduce to the legal minimum the number of people working shifts. Most of those who were taken off shifts have been found other work within the plant but some have been retrained to fulfil new staff needs.

The decommissioning of equipment no longer required for the safety of the plant, thereby reducing costs, has encouraged a change in approach by personnel, bringing to the fore new realities. Moreover, preparation of the documentation has in itself refocused the personnel involved and their abilities are now very useful for the continuing dismantling activities. Other areas of staff activity are described below.

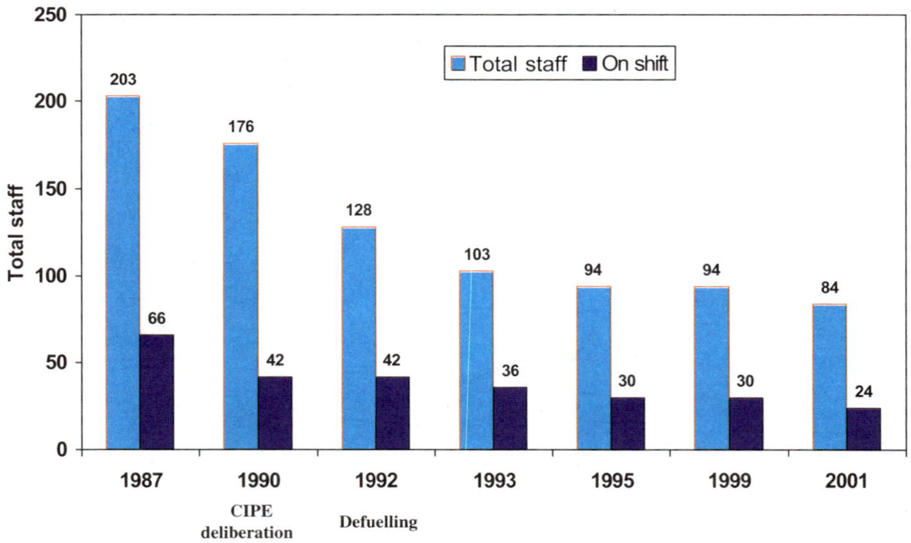

FIG. I–20. Staff numbers at the Trino NPP.

(a) *Chemistry and health physics department staff*

(1) Some staff have been used to provide radiological protection support to conventional plants which have instruments containing radioactive sources such as smoke detectors, instruments for measuring atmospheric dust, and gas chromatograph instruments in chemical laboratories (Caorso).

(2) A calibration centre for radiological instruments, which is a member of Italy's calibration service, has been developed at the Caorso NPP. All the radiological instruments of the four Italian NPPs are now calibrated here and the centre provides services to external customers.

(3) The staff perform whole body monitoring for local universities and other firms.

(4) A radiological characterization model of the plant is being developed as an integral part of the planning for further dismantling.

(5) The staff characterize and treat radioactive wastes produced during the operating period (for example by supercompaction).

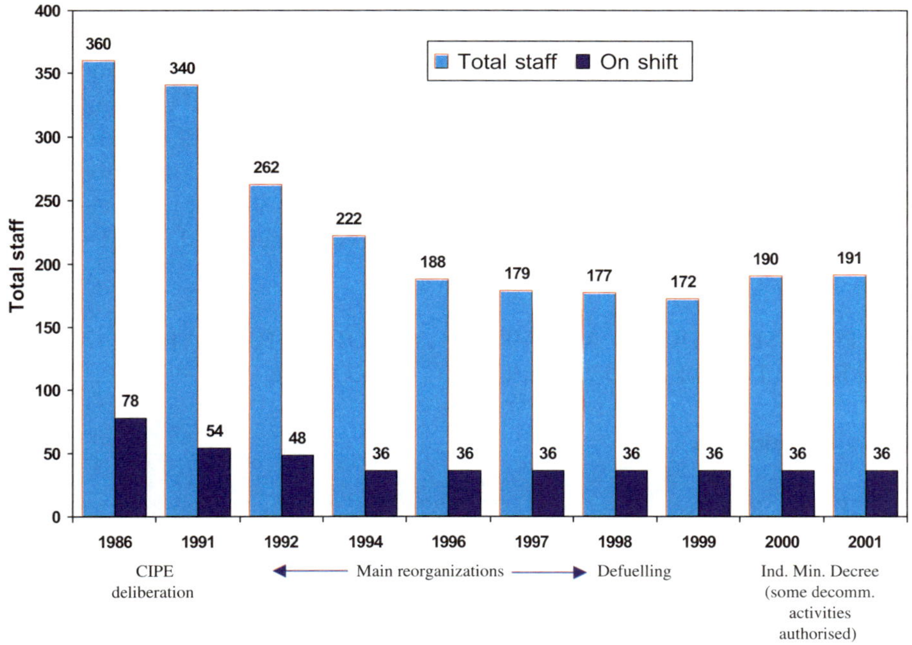

FIG. I–21. Staff numbers at the Caorso NPP.

(b) *Maintenance department staff*

The company created working teams of specialists operating in all its NPPs. For example teams of mechanics from the Caorso NPP retrieved waste from tanks at the Garigliano NPP; others worked on the retirement of systems or reconfiguration at Trino.

(c) *Operations department staff*

Operations department staff provided technical advice abroad. Staff from Caorso and Trino supported the startup of the Laguna Verde NPP in Mexico which is a BWR similar to Caorso. The department staff also supported technical staff in planning decommissioning activities.

I–6.4. Conclusions

The transition phase between operation and decommissioning in Italy revealed considerable uncertainties in both the timescale and the strategy. Nevertheless several activities carried out within the scope of the technical requirements and the operating licence have been carried out satisfactorily. Their main objectives were risk reduction, removal of dangerous substances, reduction of operating and maintenance costs, personnel training for future decommissioning activities and accurate radiological characterization of the plant and operating waste. All these activities have contributed to maintaining job motivation and took advantage of the facility's operational resources and personnel knowledge before they were lost. During the transition period some specific problems, e.g. implementation of new methods for radiochemical analysis and dismantling of special components, have been tackled and solved.

I–7. POST-SHUTDOWN ACTIVITIES AT KOREA RESEARCH REACTORS 1 AND 2, REPUBLIC OF KOREA

I–7.1. Introduction

Korea Research Reactor 1 (KRR-1) and Korea Research Reactor 2 (KRR-2) were both shut down in 1995. Korea Research Reactor 1, the first research reactor in the Republic of Korea, had been in operation since 1962, and KRR-2 since 1972. A new and more powerful research reactor, the high-flux advanced neutron application reactor (HANARO) is now in operation at the Korea Atomic Energy Research Institute (KAERI) in Daejon. Both

KRR-1 and KRR-2 are self-contained open pool TRIGA type reactors situated in tanks filled with cooling water. Korea Research Reactor 1 is a TRIGA Mark II which first went critical in May 1962 and could operate at power levels of up to 250 kW. Korea Research Reactor 2 is a TRIGA Mark III which could operate at power levels up to 2 MW.

The KRR-1 and KRR-2 decontamination and decommissioning (D&D) project commenced in January 1997 and is scheduled to be completed in 2008 when a repository for the disposal of low and intermediate level waste will be in operation. The Korea Atomic Energy Research Institute, the owner of KRR-1 and KRR-2, submitted the decommissioning plan and environmental impact assessment report to the Ministry of Science and Technology (MOST) for a decommissioning licence in December 1998. This was approved in November 2000 after long consultation and debate by the Radiation Protection Sub-Committee on Nuclear Safety and by the Nuclear Safety Commission, the highest level commission dealing with radiation safety issues in the Republic of Korea.

I–7.2. Overview of the installations

The general characteristics of KRR-1 and KRR-2 are summarized in Table I–IV. The main differences between KRR-1 and KRR-2 are the types of reactor core (fixed core versus movable core), thermal capacities (250 kW versus 2 MW), the reflector type (graphite versus H_2O) and the enrichment of the ^{235}U (20% versus 70%). In both reactors the fuel consists of a solid homogeneous mixture of uranium–zirconium hydride alloy containing 8.5% by weight of enriched uranium.

KRR-1 was initially designed and constructed in 1962 to provide up to 100 kW of thermal power, but was upgraded to 250 kW in 1969. Both KRR-1 and KRR-2 were used for research into reactor characteristics, material tests, texture studies, radioisotope production, training, education and other basic research activities involving the use of neutron beams.

I–7.3. Decommissioning licensing requirements

To obtain a licence from the Government (MOST), a decommissioning plan, including an environmental impact assessment, must be prepared according to Article No. 31 of the Atomic Energy Act. The plan must cover:

(a) The decommissioning method and schedule,
(b) Decontamination methods,
(c) Treatment and disposal of radioactive wastes,

TABLE I–IV. CHARACTERISTICS OF KRR-1 AND KRR-2

Characteristics	KRR-1	KRR-2
Reactor type	Open pool, fixed core	Open pool, movable core
Thermal power (kW)	250	2000
First criticality	3 March 1962	10 May 1972
Shutdown	January 1995	December 1995
Total operating time (h)	36 000	55 000
Total generating power (MW·h)	3 700	69 000
Neutron flux ($n \cdot cm^{-2} \cdot 5^{-1}$)	1×10^{13}	7×10^{13}
Fuel		
Contents of U (w/o)	8.5	8.5
Enrichment (w/o)	20	70
Cladding	Al	304SS
Chemical composition	$U\text{-}ZrH_{1.0}$	$Er\text{-}U\text{-}ZrH_{0.6}$
Moderator	H_2O	H_2O
Coolant	H_2O	H_2O
Reflector	Graphite	H_2O
Control rod	B_4C	B_4C

(d) Radiation protection countermeasures,
(e) Environmental impact assessments and countermeasures,
(f) Other matters as required by the MOST.

An environmental monitoring plan, covering activities prior to and during the decommissioning work, must be included in the environmental impact assessment report, in accordance with (MOST) Ministry Ordinance No. 96-31. The effectiveness of the environmental monitoring plan will be checked throughout the project and regular assessment reports forwarded to MOST. The purpose of these assessments is to ensure that the surrounding environment is not affected by the decommissioning activities, to determine what radioactive materials are accumulating in the environment and to gain the confidence of the general public by the regular promulgation of monitoring results.

I–7.4. Description of the transition period for KRR-1 and KRR-2

I–7.4.1. Spent fuel removal

The spent fuel from KRR-1 and KRR-2 is stored in the HANARO spent fuel pool in Daejon and in the KRR-2 reactor pool in Seoul. Before the D&D work was started all spent fuel was removed from the reactor site. In the meantime the US Government initiated a policy allowing for the return of foreign research reactor fuel to the USA. The March 1996 'record of decision' re-established the return programme for spent fuel enriched in the USA. An inspection team from the US DOE examined all the spent fuel from KRR-1 and KRR-2 in Seoul and Daejon during May and June 1997. The Government of the Republic of Korea consented to the policy and preparatory work for the shipment of fuel from both sites in June 1998. All the spent fuel from KRR-1 and KRR-2 was safely transported to the USA in July 1998.

I–7.4.2. Radiological characterization

The residual radiation and radioactivity levels in KRR-1 and KRR-2 have been measured, analysed and evaluated to establish and provide the technical requirements for the safe decommissioning of these facilities including minimizing radiation exposure to the workers and preventing the release of radioactive materials to the environment. The radiation dose rate and surface radioactivity contamination levels on the walls, floors and experimental equipment used within the facility were measured and evaluated. The degree of activated materials within the reactor pool structures and the levels of radioactivity and radionuclide inventory in the pool and cooling water were also analysed. Measurement and assessment of the activated reactor pool structures proved very difficult and the ORIGEN computer code was used to estimate the levels of residual radioactivity.

I–7.4.3. Decommissioning design

The design work for D&D of KRR-1 and KRR-2 was carried out in 1998. This involved thorough assessment and analysis of the design details of the reactors using operating experience, construction diagrams and radiological survey reports. Following the pre-design, hazard and operability studies were carried out for each planned working step during the D&D. One of the most significant problems was determining the radioactivity levels in the rotary specimen rack (RSR) which is located around the reactor core and is composed of aluminium alloy and a small number of stainless steel components (~3.4 kg).

The ORIGEN 2.1 computer code was used to calculate estimated dose rates. It is anticipated that these RSRs (two in KRR-1 and one in KRR-2) will be classified as intermediate level radioactive waste. As a graphics tool, the project used a commercial software package, IGRIP (Interactive Graphics Robot Instruction Program). This program is capable of performing real time graphic animation and computation. The graphic simulation system is composed of a facilitiy and equipment modelling program, a simulation program and a program allowing connections to external equipment. The graphic model of the reactor is illustrated in Fig. I–22 [I–4].

I–7.4.4. Radiation protection and health physics

Radiation protection in the decommissioning plan is one of the most important issues to be considered to minimize radiation exposure to workers and prevent the release of radioactive materials to the environment. Control of the restricted area and the necessary radiation protection facilities will require continued update and amendment of the plan as the D&D project progresses in order to provide effective radiation control and physical protection as

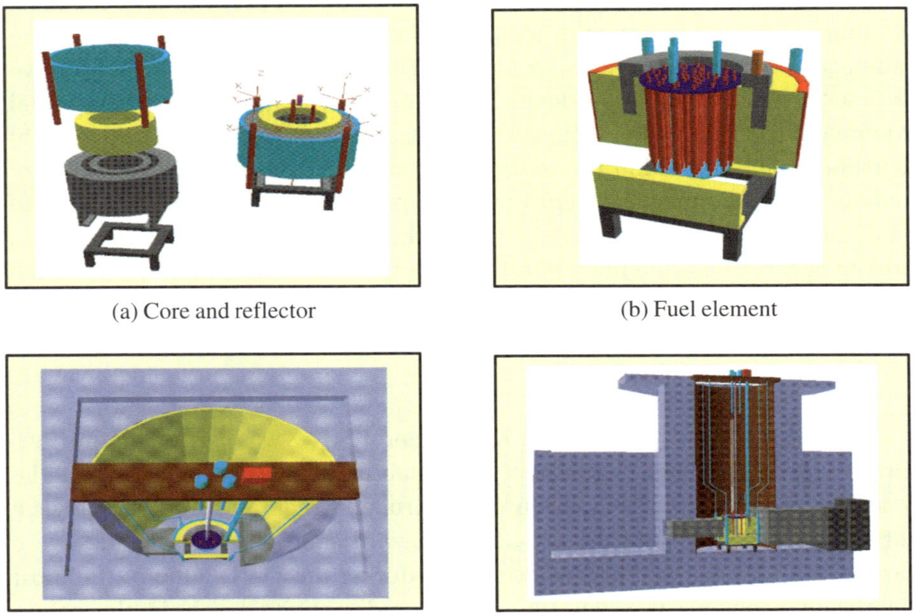

(a) Core and reflector (b) Fuel element

(c) Centre channel (d) Reactor internals

FIG. I–22. Graphic model of a TRIGA research reactor (Republic of Korea) by IGRIP.

required by the movement of human and material resources. Suitable radiation measuring equipment was prepared in advance and put in place to survey contaminated and activated materials as the work progressed. The site release criteria for unrestricted use was set at 0.4 Bq/cm^2 (or Bq/g) for β and γ emitters and 0.04 Bq/cm^2 (or Bq/g) for α emitters. A project specific annual dose limit of 15 mSv was established. All external and internal doses will follow the ALARA principle. Documents such as the radiological control manual, radiation safety control procedure, emergency procedure and training programme were prepared before physical decommissioning work commenced. Radiation work permits and access control equipment will be used to manage exit and entry procedures in the working areas. Radiological surveillance of all working areas will be performed throughout the project to ensure that the prevailing conditions are known and that the appropriate controls are implemented. Monitoring will be implemented to limit unnecessary exposures and prevent the spread of contamination.

I–7.4.5. Preparatory work

Preparatory work included the installation of radiation measuring and analysis equipment, a turnstile access gate equipped with an automatic individual radiation exposure recording system, hot and cold shower rooms for the workers, laundry equipment and the installation of radioactive liquid waste treatment facilities such as a natural evaporator and membrane equipment for the liquid wastes from the laundry and showers.

I–7.5. Conclusions

The decommissioning of the KRR-1 and KRR-2 reactors, the first experience of its kind in the Republic of Korea, commenced at a time when there were no existing or well established D&D procedures or regulations. The necessary regulatory regime was subsequently established both through post-shutdown activities and ongoing licensing work. These activities and assessments identified the necessary technology and equipment needed for the dismantling work together with the waste management plans and relevant exemption levels. In the future, ongoing and regular radiological surveys will be carried out on all facilities and sites prior to D&D to determine the exact volume of radioactive waste to be managed. The post-shutdown activities on the KRR-1 and KRR-2 reactors included removal of the spent fuel (unloading and transportation), radiological characterization, decommissioning design and preparatory work.

I–8. POST-SHUTDOWN ACTIVITIES AT THE DODEWAARD NPP, NETHERLANDS

I–8.1. Introduction

I–8.1.1. Dodewaard NPP

The Gemeenschappelijke Kernenergiecentrale Nederland (NV GKN) is the owner and operating organization of the Dodewaard NPP (KCD), the first NPP in the Netherlands. The KCD is a small boiling water reactor with natural circulation. Its mission was to gain construction and operating experience with nuclear energy for electricity generation and create research opportunities in the field of nuclear energy. Its main characteristics are listed in Table I–V.

When it became clear in 1996 that the construction of new NPPs in the Netherlands was not feasible in the short term, the Dodewaard NPP was shut down. Electricity production ceased on 26 March 1997, exactly 28 years after the power plant had officially been put into operation. Now that production has stopped, the plant is preparing for a 40 year SE period. The SE includes all of the measures necessary to keep radioactivity within a defined and controlled area. Final dismantling is expected to start in 2043.

I–8.1.2. The post-operational period

The post-operational period is divided into three phases:

- Phase 1: fuel in the reactor vessel and the fuel pool.
- Phase 2: fuel in the fuel pool and/or partially disposed of.
- Phase 3: fuel disposal completed.

All activities carried out during the three phases are intended to ensure that post-shutdown operations are safe, responsible and efficient. Activities include retiring non-essential systems and the continued operation, supervision, testing and maintenance of essential systems. A system is defined as non-essential if it has no further function in the current or subsequent phases.

Phase 1

Phase 1 commenced in March 1997 when electricity production ceased and continued until all the fuel had been removed from the reactor core and placed in the fuel pool. In this phase, while the reactor core was being emptied, the safety level was similar to that during a scheduled major maintenance stop.

TABLE I–V. MAIN CHARACTERISTICS OF KCD

Reactor type	BWR with natural circulation
Thermal capacity	183 MW(th)
Electrical capacity	58 MW(e)
Average load factor during the plant's operational life	84.1%
Start of construction of the Dodewaard NPP	May 1965
Official start of operation	26 March 1969
Last day on the grid	26 March 1997
Post-operational period	March 1997–2002 (due to fuel transports)
Planned preparation for SE	2003–2004
Planned safe enclosure period	2005–2042
Planned final dismantling	2043–2047

Phase 2

Phase 2 began in October 1997 and will continue until all the fuel in the fuel pool has been removed from the site. Following fuel removal, the nuclear risk to the public will be negligible as reactivity accidents will not be possible. A new licence for preparing an SE period of approximately forty years will be required before final dismantling takes place.

Phase 3

Phase 3 will begin once all the fuel has been removed from the site. There will be no nuclear risk to the public during this phase. While the overall risk to plant staff will diminish with time, a small radiological risk will remain owing to the presence of radioactive materials on the site.

I–8.2. Operational evaluation

Operational evaluation is divided into five parts which cover all relevant issues:

(1) *Part A: buildings, installations and systems*

- Use of systems,
- Maintenance,
- Surveillance,
- Taking systems out of service,
- Control of modifications,
- Operation/monitoring (shift occupation, emergency operating procedures),
- Redundancy,
- Technical specifications (in part).

(2) *Part B: control of radioactivity*

- Releases to the atmosphere and river,
- Monitoring air and water quality,
- Radiological protection,
- Technical specifications (in part).

(3) *Part C: disposal of radioactive materials*

- Processing of radioactive waste,
- Disposal of fuel elements (both irradiated and non-irradiated),
- Technical specifications (in part).

(4) *Part D: planning, recording, evaluation and reporting*

- Planning (weekly, monthly, etc.),
- Recording of operational data (including monthly reports),
- Recording, evaluation and reporting of internal faults,
- Consideration of external faults,
- Probabilistic safety analysis (PSA),
- Information technology,
- Technical specifications (in part).

(5) *Part E: other issues and general matters*

- Training;
- Environmental management;
- Occupational health and safety;

- Emergency plan, internal emergency services, first aid in accident situations and consignment;
- Insurance;
- Subscriptions;
- Control of documents;
- Public relations;
- Security.

Matrix structure evaluation normally takes place by combining information from Parts A to E relative to the three generic phases.

I–8.3. Description of the process, activities and tasks

I–8.3.1. Phase 1: fuel in the reactor vessel

During this phase, the focus is towards controlling the reactivity risk from the fuel remaining in the reactor. However, since electricity generation has ceased, certain non-essential systems can be taken out of service and retired.

I–8.3.1.1. Part A issues: buildings, installations and systems

System retirement. System retirement entails shutting systems down and modifying the regular maintenance and surveillance programmes accordingly. As a result of the reduced maintenance and surveillance, a system is no longer considered qualified for its original purpose once it has been retired. Figures I–23 and I–24 provide examples of such systems.

Many systems can be declared non-essential during phase 1, including nuclear safety-related systems, reactor systems, all turbine related systems and specific water quality control systems. Procedures have been developed for retiring non-essential systems based on technical scenarios or questions such as:

(a) Which other systems are relevant in relation to taking the system in question out of service?
(b) How is the system in question to be stored (dry, wet, inert, etc.)?
(c) How is the system to be flushed out, cleaned and decontaminated for subsequent dry storage without creating a hazard for personnel or the environment (e.g. through air contamination)?
(d) Which procedures or checklists should be retained, modified or drawn up (with regard to safe and effective working, data registration, comparison of data against prescribed criteria, etc.)?

(e) Which procedures and/or checklists are no longer relevant?
(f) How are the relevant documents to be archived?
(g) Who will carry out the activities, when and in what order?
(h) Which permits need to remain in force (such as steam systems that may be covered by statutory steam equipment regulations)?

Maintenance. Systems that have not yet been taken out of service will be maintained in the normal manner. Once a system has been retired it will receive only the minimum amount of maintenance necessary to preserve its physical integrity pending subsequent (definitive) decommissioning and conservation. This maintenance will consist of activities such as mechanical integrity inspections and corrosion inspections of hangers, pipe work, etc.

Surveillance. Surveillance regimes will be similar to those used during normal operation. No new specific age monitoring or management systems will be developed. Surveillance will continue to include:

FIG. I–23. The control room of the Dodewaard NPP with inactive panels indicating that the systems are out of operation.

(1) Functional tests,
(2) Monitoring of air and water quality,
(3) Periodic preventive maintenance,
(4) In-service inspections.

Checklists and procedures for systems that have yet to be taken out of service will be applied and updated in the normal manner.

Control of modifications. In principle, once the reactor has been shut down, formal modifications will no longer be required. Any changes necessary to systems relevant to safety will be made on an ad hoc basis, using a simplified version of the existing modification procedure for guidance. The nuclear inspectorate must be notified of any such changes.

Operation/monitoring. During this phase a nominal staff of five shift personnel will be maintained, with four at a minimum. All emergency operating procedures will remain in force during phase 1. During this phase existing technical specifications will be supplemented by a redundancy plan, as was normally done during previous operational refuelling stops.

FIG. I–24. Plugged valves of systems taken out of operation at the Dodewaard NPP.

I–8.3.1.2. Part B issues: control of radioactivity

Releases to the atmosphere and river. Following shutdown of the reactor and while fuel still remains in the core the normal amount of operational radioactive gaseous material present will be greatly reduced. Therefore, less emphasis needs to be placed on the management of gaseous release levels. However, since the possibility of fission product release remains, particularly during accident scenarios, air monitoring will continue to be required.

During phase 1 the amount of radioactive material associated with the processing of wastewater will remain almost unchanged. While a small number of systems will be washed out and stored dry, this will not significantly increase the amount of wastewater produced. Nevertheless, suitable water quality discharge procedures will remain in force to both clean up contaminated water and allow clean water to be released into the river.

Monitoring air and water quality. Monitoring of air and water quality will continue during phase 1. Particular emphasis will be placed on air quality, in view of the possibility of air contamination by aerosols, since a small number of systems will be washed out and left dry. The water quality standards normally associated with a prolonged stoppage or outage will continue to apply.

Radiological protection. Only minimal changes will be made to existing radiological protection procedures, supervisory requirements and expertise needed during phase 1. Although there will be fewer areas with relatively high radiation dose rates such areas will still need to be measured, delineated and controlled. Appropriate measures will be taken to maintain the ALARA principle, including environmental monitoring and reporting.

I–8.3.1.3. Part C issues: disposal of radioactive materials

Processing of radioactive waste. During phase 1 most systems that make use of filter material and/or resins (ion exchangers) will remain operational. Hence the supply and usage of such materials, particularly in the wastewater storage tank areas, will not be significantly greater than during normal operation.

If wastewater cannot be discharged into the River Waal it will be evaporated in the wastewater evaporation system and the resultant brine concentrate will be transferred to the storage tanks. The waste collected in these tanks will then be conditioned by cementation during periodic waste campaigns. Processing will be done in the usual way, using the normal procedures and checklists. Waste processed in this way will be temporarily stored in the waste building or elsewhere within the controlled area prior to forwarding to COVRA (Centrale Organisatie voor Radioactief Afval N.V.).

During phase 1, the amount of 'domestic' radioactive waste produced (work clothing, packaging, contaminated components, etc.) will not be greater than during normal operation. There will be no increase in the amount of installed or core component radioactive waste stored in the fuel pool apart from the fuel elements themselves. The fuel element linings will also be kept together in the fuel pool, thereby creating extra workspace for loading the fuel transport container. Tools and equipment not needed in subsequent phases will be cut up into small pieces/lengths (~60 cm) and will be compressed and packed in 100 litre transport drums for forwarding to COVRA. Small instrument calibration radiation sources will still be needed and will not be disposed of at this stage.

During phase 1, all waste processing activities will be comprehensively optimized. This will include a review of all existing procedures and checklists relating to waste processing.

Disposal of irradiated fuel. During phase 1 shift personnel will take the 164 fuel elements from the reactor vessel and place them in the fuel pool. In principle, the transfer could begin as soon as the reactor water temperature has fallen to below 60°C, which is anticipated to take about five days. It is not anticipated that any irradiated elements and/or fuel rods will be removed from the power plant during this phase.

Disposal of non-irradiated fuel. The non-irradiated fuel elements, fuel rods and reference fuel tablets stored in the dry fuel storage facility can be disposed of in phase 1. The transfer of the materials to a suitable fuel processing organization will be the responsibility of the fuel management and contracts group.

I–8.3.1.4. Part D issues: planning, recording, evaluation and reporting

Planning. Several of the existing planning regimes will lose their relevance during phase 1, including planning for an annual refuelling and maintenance stop and quarterly test planning. Other regimes, such as weekly planning, system stop planning and steam equipment service inspection planning will remain in force.

During all three phases:

(a) The planning regimes will cover a decreasing number of systems,
(b) There will be decreases in the amount of preventive maintenance and the number of modifications required,
(c) There will be changes in the staffing arrangements,
(d) There will be increases in planned activities associated with taking systems and/or installation components out of service.

Because of the constant changes and the increasingly interdependent nature of the activities, there will be a greater need for strict, centralized co-ordination. A centralized system of planning and recording covering activities, staffing and work progress will have to be prepared and, if possible, implemented in phase 1. This system will supersede the individual group planning regimes.

Recording of operational data (including monthly reports). The recording and monthly reporting of relevant operational data will continue in phase 1. The number of reports on tests, modifications, changes and faults is expected to decrease, due to the fact that the reactor is shut down and the requirement for major installation modifications will have ceased. On the other hand, the monthly reporting system will be extended to include matters relating to new circumstances, e.g. future monthly reports will include a section on systems taken out of service.

Probabilistic safety assessment (PSA). A comprehensive live PSA will not be required.

I–8.3.1.5. Part E issues: other issues and general matters

Training. Training of the qualified reactor operators and other shift personnel will continue, with the emphasis changing to reflect both the operational situation and taking systems out of service. A refresher training programme will remain in place, albeit in a modified form.

For managers the focus will shift to courses on leadership and communication skills. The training made available to other personnel has been supplemented by courses on communications skills, holding of effective meetings, leadership, management, safety and specialized job related courses where relevant. Courses on reactor technology will no longer be given.

Environmental management. There will be no policy changes in this area.

Occupational health and safety. There will be no policy changes in this area.

Emergency plan, internal emergency services and first aid in accident situations. Although the external risk levels will be negligible about 150 days after final shutdown, there will be no changes made to the emergency plan or the internal emergency services and an assignment system will remain in effect.

Insurance. As the risk decreases and the number of operational systems declines, the various insurance policies can be modified.

Document control. There will be no procedural changes, but the archive will be rationalized. No archived material will be disposed of until a review has been undertaken to determine which data are likely to be required in the future. An inventory of all existing archives will have to be made.

Public relations. The operational dissemination of public information will cease but presentations may still be made to business contacts. The information and materials will be kept up to date and the information facilities will be retained.

Security. There will be no policy changes in this area.

I–8.3.2. Phase 2: fuel in the fuel pool

During this phase all fuel will be in the fuel pool ready for disposal. Additional systems, including certain safety related systems, can then be retired. Among these are the core emergency inundation system and the reactor protection system. Operational activities will be adjusted accordingly. The waste processing programme initiated during phase 1 will continue during phase 2. Issues associated with phase 2 activities include:

Disposal of irradiated fuel. The 164 fuel elements from the last reactor cycle need to be cooled for at least 180 days before they can be loaded into a transport container for shipment to the reprocessing facility.

Disposal of non-irradiated fuel. Any remaining non-irradiated fuel will be disposed of by shipment to the fuel manufacturer.

Liquid process waste. The processing and shipment of conditioned liquid process waste to COVRA will continue. This is one of the major activities during phase 2.

Solid low level radioactive waste. The selection, collection, cutting, packaging and shipment of relatively small quantities of solid low level radioactive scrap material to COVRA will continue during this phase. Alternative disposal routes have been considered but have not turned out to be cost effective.

Solid intermediate level radioactive waste. The solid waste (other than fuel) in the reactor spent fuel pool will be characterized in detail. A disposal programme will then be developed for subsequent execution in phase 3.

I–8.3.3. Phase 3: fuel disposal completed

Phase 3 will commence when fuel disposal has been completed. The licence for preparing the SE will not yet be in force. A few more systems, such as the pool cleaning system, may be retired during this phase but modifications needed for conservation of the plant during the SE waiting period will not be able to be installed. One of the major issues associated with phase 3 activities includes the disposal of radioactive materials.

Disposal of radioactive materials. It is GKN's policy to dispose of all liquid and solid radioactive waste before the beginning of the SE period. Consequently the waste removal programme begun during phase 2 will be continued and completed.

As a result of the decreased levels of activity within the controlled area, the flow of solid 'domestic' radioactive waste (work clothing, packaging, contaminated components, etc.) will gradually decline during this phase. This waste will be sorted, compressed and then forwarded to COVRA as in phase 2.

I–8.4. Conclusions

The post-shutdown activities at the Dodewaard NPP focused primarily on the timely achievement of safe and stable facility and management operations during what proved to be a very hectic phase of the plant's lifetime. All aspects of plant operation (technical, operational, administrative and personnel) were considered and reviewed. The review indicated that many process systems could be retired soon after final shutdown of the plant, depending on the spent fuel situation, thereby considerably reducing the maintenance and surveillance effort.

The shipment of all spent fuel from the plant will further decrease the risk to personnel and the environment. The management of radioactive waste is a major task involving not only standard process wastes but also other radioactive wastes such as solid low level radioactive waste and intermediate level radioactive waste stored in the spent fuel pool.

Careful attention has been paid to preventing accidents and incidents which could lead to inadvertent releases into the environment during the many non-standard operations that will be carried out during post-shutdown activities. Air and water quality monitoring will continue during this period, with particular emphasis on air quality, in view of the possibility of air contamination by aerosols.

Particular attention was paid to the management and organizational aspects. The actual size of the organization at any given time must be commensurate with both immediate and future needs and activities. This aspect requires careful planning, preferably carried out prior to final shutdown. As was anticipated, the radiological protection organization has remained almost unchanged from that during normal plant operation.

I–9. THE TRANSITION PERIOD IN SLOVAKIA

I–9.1. Introduction

Post-shutdown and pre-decommissioning planning activities in Slovakia reflect the fact that the first 'standard' shutdown of an NPP is planned for 2006. Based on experience and the lessons learned from decommissioning of the A-1 reactor, which was shut down in 1997 on the order of the regulatory authority after an accident, the primary aim was the preparation of conceptual decommissioning plans for all nuclear installations and more detailed decommissioning documentation for the V-1 NPP.

Decommissioning related documentation produced to date includes detailed technical reports on post-shutdown system reductions, modifications, conceptual decommissioning plans, periodic updates, reuse of equipment, the buildings and the site, environmental assessments, funding, social and economic impacts.

I–9.2. Nuclear facilities and installations

The major nuclear facilities and installations in Slovakia are:

(a) The A-1 reactor at Bohunice, a natural uranium fuelled, heavy water moderated and CO_2 cooled reactor. It was shut down in February 1977 and the decision to decommission it was taken in 1979. Spent fuel was removed in 1999 and decommissioning activities are ongoing with the aim of achieving SE in 2007.

(b) The V-1 NPP at Bohunice. Two 440 MW WWER-440, V-230 reactor units were commissioned in 1978 and 1980. Both units were progressively upgraded from 1993 to 2000 with the main aim of increasing nuclear safety. Closure of both units is planned for 2006 and 2008.

(c) The V-2 NPP at Bohunice. Two 440 MW WWER-440, V-213 reactor units were commissioned in 1984 and 1985. Their design lifetime is 30 years and their closure is planned for 2014 and 2015.

(d) The Mochovce NPP. Two 440 MW WWER-440 V-213 reactor units were commissioned in 1998 and 2000. Their closure is planned for 2028 and 2030, respectively. Construction of two other units of the same type has been put on hold for the time being.

I–9.3. Licensing requirements

The licensing requirements and legislative background of the transition period are detailed in the following acts and decrees:

(a) Act No. 130/1998 Col. of the National Council of the Slovak Republic on the Peaceful Use of Nuclear Energy (commonly referred to as the 'Atomic Law'). The Nuclear Regulatory Authority of the Slovak Republic (ÚJD SR), as a central body of the state administration, carries out state supervision during the design, construction, commissioning, operation and decommissioning of all nuclear installations in accordance with this act and its relevant regulations. Article 19 of the law, which covers decommissioning related documentation, is established by Decree 246/1999 which is issued by the ÚJD SR.

(b) Decree 246/1999. Decommissioning documentation for nuclear facilities. This decree defines the scope and contents of the decommissioning documentation that has to be submitted to the ÚJD SR by the operating organization. The types of documentation that needs to be submitted to the ÚJD SR encompass:

   (1) Approval of the following documents:
   - Technical specifications,
   - Quality assurance programmes covering actual decommissioning or the phases of decommissioning,
   - On-site emergency plans.

   (2) Evaluation of the following documents:
   - Decommissioning plan or plans for particular phases,
   - Decommissioning strategy/concept for the period immediately following the authorized decommissioning phases,
   - Physical protection plans,
   - Radioactive waste management systems,
   - Off-site radiation control programmes,
   - Equipment inspection programmes,
   - Selected operating procedures,
   - Professional qualification documents,
   - Insurance policies or other documents detailing financial liability.

In addition to these main regulations, various acts and decrees deal with other aspects of the decommissioning transition period:

(c) Act No. 254/1994 Col. of the National Council of the Slovak Republic on the State fund for decommissioning of nuclear power installations (as changed and amended by Act No. 78/2000). This is the state fund for nuclear power facilities, decommissioning, spent nuclear fuel and radioactive waste management.
(d) Public Notice No. 14/1995. Details of contributions, granting and use of funds.
(e) Act No. 127/1994. Environmental impact assessments.
(f) Decree No. 190/2000. Regulation and management of radioactive waste and spent fuel.
(g) Decree No. 12/2000. Radiation protection, including decommissioning.
(h) Act No. 272/1994 (as changed and amended by Act No. 470/2000). Site release criteria, material clearance and authorized release, in accordance with IAEA/NEA guidance [I–5].

I–9.4. Decommissioning strategy

According to the Atomic Law, the decommissioning strategy is to be developed by the operating organization of a nuclear facility, taking into account the actual legislative framework for decommissioning together with the existing and planned decommissioning infrastructure. This strategy defines the operating organization's structure for subsequently working up the decommissioning options, which should cover all anticipated decommissioning activities. The most important factors for developing the decommissioning strategy are:

(a) The actual and anticipated state of the plant at the time of decommissioning;
(b) The actual or planned state of repositories for the final disposal of conditioned radioactive waste;
(c) Available and planned methods and systems for the treatment, conditioning, storage and transportation of radioactive waste;
(d) Material release systems and arrangements;
(e) Availability of funding;
(f) Availability of technical equipment, procedures and qualified personnel;
(g) Environmental aspects;
(h) Social and economic aspects.

An extensive technical decommissioning infrastructure has already been established in Slovakia to deal with the unplanned decommissioning of the A-1 reactor. A specialized subsidiary of Slovak Electric (SE VYZ), responsible for decommissioning, spent fuel management and radioactive waste management, was established in 1996 to operate the following licensed facilities and systems:

(1) Interim spent fuel storage facilities for WWER type NPPs.
(2) Radioactive waste treatment facilities, including systems for fragmentation, decontamination, cementation, bituminization, super-compaction, incineration and final conditioning into reinforced concrete containers.
(3) Interim storage facilities for conditioned and non-conditioned radioactive waste. These will be further extended, particularly for conditioned waste intended for future disposal in a geological repository.
(4) A low and medium level radioactive waste repository has been established at Mochovce and was granted a licence in 2001.
(5) A future geological high level radioactive waste repository is planned and is expected to be ready for use some time after 2037.

The conceptual plans for decommissioning NPPs in Slovakia cover three main options:

(i) Immediate dismantling,
(ii) Safe enclosure,
(iii) Enclosure with surveillance.

Valuable experience has been gained and a pool of experienced and qualified personnel has been established during the ongoing decommissioning of the A-1 reactor, primarily by experienced contractors who have provided project and engineering services (decommissioning documentation, cost estimates, research and development), technical services (decontamination and refurbishments), design services and specialized equipment, as shown in Fig. I–25.

I–9.5. The transition period

In Slovakia, the studies carried out for the transition period from NPP operation to decommissioning can be divided into two stages:

(1) The period up to 1998 when the Atomic Law was issued,
(2) The period following the issue of the Atomic Law.

Until the Atomic Law was issued the lessons learned during the premature shutdown of the A-1 reactor and the subsequent preparations for decommissioning were the main driving force for thorough, timely and appropriate studies and for the production of decommissioning documents for the WWER NPPs, including the development of an appropriate decommissioning strategy for the V-1 NPP. It should be noted that the development of such documentation was not required by any existing regulations before the Atomic Law was issued. When the Atomic Law and related regulations were issued, conceptual decommissioning plans for all nuclear installations were developed and submitted to the Nuclear Regulatory Authority.

I–9.5.1. Approach to the decommissioning transition period before issue of the Atomic Law

During this period, the general studies carried out and the documents produced included:

(a) Two feasibility studies for the V-1 NPP to select the preferred decommissioning option,
(b) Preparation of relevant documents for the 'termination of operations' of the V-1 NPP,
(c) Other decommissioning related feasibility studies.

FIG. I–25. Graphic simulation (by EUCLID) of dismantling activities at the reactor.

I–9.5.1.1. Feasibility studies for the V-1 NPP

The following five decommissioning options were analysed in detail:

(1) Immediate total dismantling of the NPP after final shutdown,
(2) Safe enclosure of the 'hermetic area' (part of the reactor building) for each individual reactor unit,
(3) Safe enclosure of the reactor shaft of each individual reactor,
(4) Safe enclosure of the entire reactor building,
(5) Closure of the NPP with surveillance (stage 1 of the former IAEA classification).

The following factors were taken into account within each option:

- The total number of person-days needed and personnel requirements,
- The amount and activity of radioactive waste,
- The amount of non-radioactive waste,
- The total collective dose equivalent,
- The environmental impact of gaseous and liquid effluents,
- The technical equipment required,
- The total costs,
- Time schedules.

For all SE options, a deferral period of 70 years was assumed before final dismantling to green field status. Multi-attribute analysis was used to assess the results and select the preferred option by appropriately weighing and scoring all relevant feasibility, safety, environmental, technical, waste, social, funding and economic aspects. This analysis indicated that safe enclosure of the hermetic area of each unit was the preferred option. This multi-attribute analysis approach will be used to assess how best to proceed with all future NPP decommissioning options in Slovakia.

I–9.5.1.2. Preparation of relevant documents for the termination of operations of the V-1 NPP

'Termination of operations' is defined as the period of time from the date of final shutdown of the NPP to the commencement of SE activities. This period is divided into two stages. The initial stage, called final shutdown, encompasses activities associated with safe reactor shutdown, fuel removal and spent fuel management up to its transport to the interim spent fuel storage facility (Fig. I–26). The second stage begins at this point and ends with the erection of the SE.

The following activities were taken into consideration for final shutdown:

(a) Drainage of systems and process fluids,
(b) Defuelling of the reactors and transfer of fuel to the unit's spent fuel storage pool,
(c) Disconnecting/isolation of unused systems and equipment,
(d) Transportation of fuel from the storage pool to the interim spent fuel storage facility,
(e) Conditioning of radioactive wastes,
(f) Decontamination of the primary circuit and other equipment as required,
(g) Amendment of operational and safety documentation to reflect the operational status of the termination of operations,
(h) Modification of operational arrangements.

While many of these operations could be carried out under the normal operating licence, specific regulatory permission would need to be obtained for tasks not already in the licence such as primary circuit decontamination, dismantling, system modifications, etc. Permission for such tasks would be through approval of documented safety cases.

FIG. I–26. Transport cask for transferring spent fuel from the V-1 and V-2 NPPs to the interim spent fuel storage facility.

Producing the decommissioning documentation for the V-1 NPP proved to be very challenging as it needed to take account of various safety upgrades such as fitted equipment, potential decommissioning requirements and potential cost escalations. Each endpoint or phase in the planned decommissioning process would normally dictate what equipment and/or operational changes were required to a large number of individual systems. Once the overall system and operational requirements were known, individual documents and step by step procedures were developed to both manage and implement the various tasks. A series of safety reports and quality assurance programmes were also produced covering, inter alia, electrical supplies, instrumentation, control and radiation monitoring systems. Procedures were not developed for dealing with non-operational systems not being used for the decommissioning process.

The range of documentation and studies undertaken include:

(1) System requirements covering the current phase, preparations for the next phase and for the SE period,
(2) Equipment conservation plans for systems to be dismantled in the later stages of decommissioning,
(3) Effects of decommissioning on the V-2 NPP,
(4) Decontamination of the primary circuit after spent fuel discharge,
(5) Reduction of the fire risk,
(6) Processing, treatment and conditioning of radioactive wastes,
(7) Utilization of spent fuel from the V-1 NPP.

I–9.5.1.3. Other decommissioning related feasibility studies

Other decommissioning related feasibility studies and documents included:

(a) Environmental impact assessments into the anticipated effects of V1 NPP decommissioning on the environment, as required by Law No.127/1994;
(b) Further decommissioning studies into the V-2 NPP and the Mochovce NPP.

I–9.5.2. Approach to the decommissioning transition period following passage of the Atomic Law

Conceptual decommissioning plans were developed covering the transition period for all nuclear facilities in accordance with the requirements of the Atomic Law.

I–9.5.2.1. Decommissioning documentation and other activities

The conceptual decommissioning plans encompassed the V-1 NPP, the interim spent fuel storage facility at Bohunice (for WWER spent fuel), the radioactive waste treatment plant, the bituminization and cementation units. The general methods used to evaluate the various decommissioning options were similar to those used during earlier studies, including:

(a) Updating the documentation for the termination of operations of the V-1 NPP. This covered a detailed description of the systems needing to be kept in operation following final NPP shutdown up to the transportation of the last fuel element into the interim spent fuel storage facility and the subsequent deactivation of operational systems to levels commensurate with the next phase of decommissioning. The documentation also addressed deactivation schedules, project timescales, drawings of the modified systems, the numbers of personnel required and a comprehensive safety analysis.
(b) Socio-economic consequences of premature shutdown of the V-1 NPP.
(c) Other research and development, and technical support requirements. For example, an Oracle based computer code was developed for calculating costs and other decommissioning parameters to enable the various decommissioning solutions to be optimized. This computer code was used to implement the Proposed Standardised List of Items for Costing Purposes issued by the IAEA, EC and OECD in 1999 [I–6].

I–9.5.2.2. Conceptual plan for decommissioning of the V-1 NPP

The three main decommissioning options considered are immediate decommissioning, SE of reactors in reactor shafts, and closure with surveillance. The updated conceptual decommissioning plan for the V-1 NPP indicated that the preferred planning option was SE, most likely with a 30 year deferral period, while not totally ruling out the immediate/direct decommissioning option. The main studies covered decommissioning documentation, termination of operations and the implementation of the actual decommissioning activities as shown in Fig. I–27.

The decommissioning related documentation and studies included:

(a) An updated conceptual decommissioning plan,
(b) Environmental impact assessment documents,
(c) Optimization of fuel removal campaigns,

(d) Systems and equipment needed for both the termination of operations and during subsequent decommissioning,
(e) The residual state of equipment at the time of dismantling,
(f) Reuse of systems following decommissioning,
(g) Reuse of buildings following decommissioning,
(h) Reuse of the site.

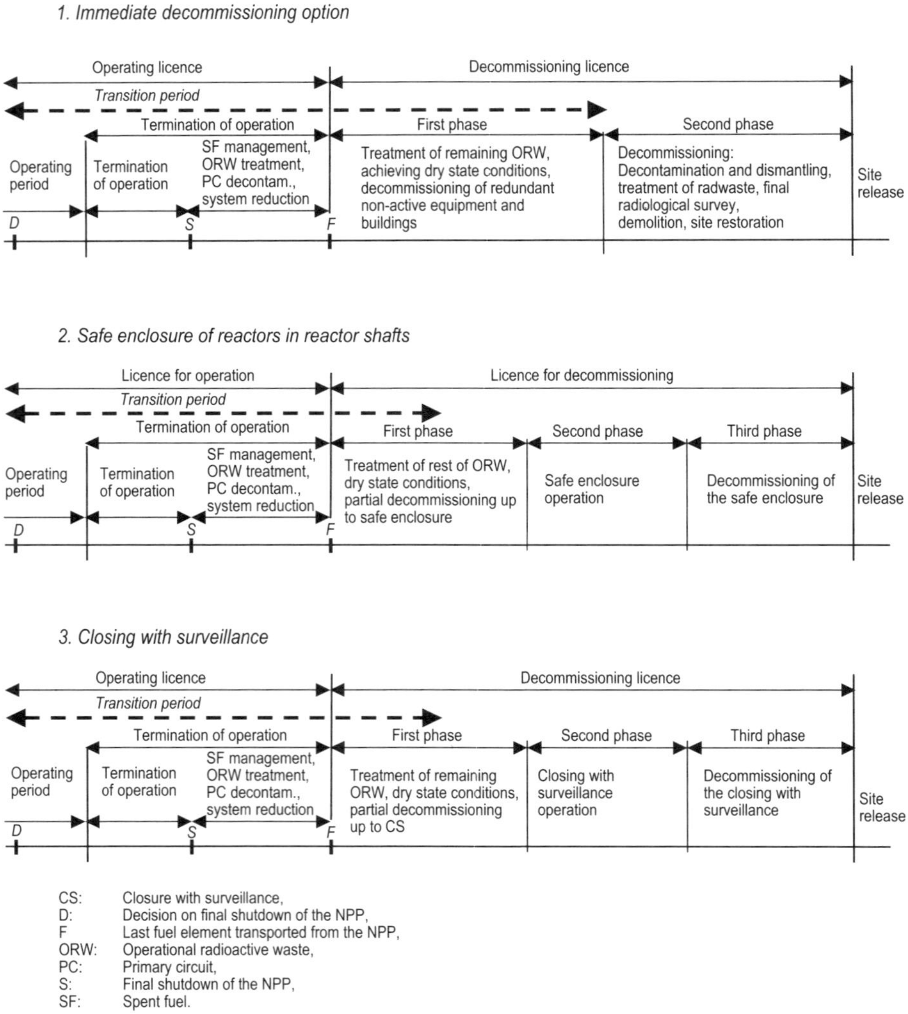

FIG. I–27. Decommissioning strategy options for the V-1 NPP.

At the end of the transition period the main documents will be:

(1) The final decommissioning plan,
(2) Project management plans for individual phases,
(3) License support documents.

Documents and studies related to the termination of operations covered:

(i) Operational training and qualification requirements up to the last day of normal operation;
(ii) Training and qualification of operating personnel required during the transition period and the decommissioning phases;
(iii) Preparation of the decommissioning database for buildings and systems;
(iv) Radiological surveys, sampling, model activation calculations and data collection;
(v) Reactor defuelling, fuel related cooling systems and transportation of spent fuel to the interim spent fuel storage facility;
(vi) Operational, combustible, toxic and radioactive waste management;
(vii) Decontamination of primary circuits;
(viii) Final shutdown inspections;
(ix) Assurance of nuclear safety;
(x) Drainage of systems, drying, reduction and conservation;
(xi) Isolation of systems and disconnection of electrical supplies;
(xii) Calculation of radioactive inventory and analysis;
(xiii) Sampling of soil and water;
(xiv) Implementation of the transition plan from operation to decommissioning.

The documents and studies covering the implementation of decommissioning, as well as the appropriate actions necessary to mitigate the consequences of the shutdown of an NPP, include:

(a) Radiation dose measurement systems and radiation protection systems during the transition period and decommissioning;
(b) Systems for monitoring the free release of materials;
(c) Modification or replacement of heating and steam supply systems;
(d) Modification or replacement of the main and alternative electrical power, distribution and communications, and media systems;
(e) Enhancement of existing radioactive waste treatment and conditioning systems;

(f) Extension of the capacity of the low and medium radioactive waste surface repository;
(g) Extension of the capacity of the interim storage facility for conditioned radioactive waste not suitable for disposal at the surface repository.

I–9.6. Conclusions

The decommissioning of the A-1 reactor, which was prematurely shut down in 1977 and defuelled in 1999, is under way with the aim of achieving a radiologically safe status in 2007. The first planned standard shutdown of an NPP in Slovakia will be the V-1 NPP at Bohunice, planned for 2006, following which decommissioning will commence.

The decommissioning options for both are still under discussion but include either a period of SE or immediate/direct decommissioning. The V-1 NPP SE deferral period is anticipated to be 30 years.

The main post-shutdown and pre-decommissioning planning studies and activities carried out to date in preparation for decommissioning have taken due account of the knowledge and experience gained from NPP decommissioning in other countries. In general terms, these studies have covered the preparation of technical documentation, safe disposal of radioactive wastes, funding and socio-economic aspects. The documentation and infrastructure required for the decommissioning transition period for the V-1 NPP is being developed alongside that of the current and future requirements for the A-1 reactor.

I–10. THE TRANSITION PERIOD IN THE VANDELLÓS 1 NPP, SPAIN

I–10.1. Background

The Vandellós 1 NPP, owned by the Hispano Francesa, S.A. company (Hifrensa), started up in 1972 and was shut down in October 1989 after 17 years of service, during which it generated an accumulated 55 647 GW·h. The plant is located on the Mediterranean coast in the municipal area of Vandellós-Hospitalet de l'Infant, some 40 kilometres from the provincial capital Tarragona. The installed electrical power of the plant amounted to 497 MW.

This NPP is the only Spanish member of the family of plants that use natural uranium as fuel, graphite as the moderator and CO_2 as the coolant. It is based on the European natural uranium–graphite–gas model developed by the UK and France, its design being based on a joint project of EDF and CEA. The Vandellós 1 NPP is a duplicate of the French Saint-Laurent-des-Eaux plant.

The plant was shut down immediately after a fire that occurred on 19 October 1989 and caused damage to the conventional installations. This resulted in the Ministry of Industry and Energy ordering the suspension of its operating permit in November 1989. The plant subsequently remained in the shutdown condition pending reports on the causes and effects of the fire and studies and technical decisions initiated with a view to its possible recovery.

In July 1990, following the analysis of these reports, the Directorate General for Energy decreed by Ministerial Order that the plant's operating permit should be definitively suspended. This order also established the conditions under which Hifrensa was to address the pre-dismantling phase in order to keep the plant in a safe shutdown condition, remove the spent fuel from the site and condition other wastes produced during operation. In accordance with this order, Hifrensa removed the spent fuel (high level radioactive waste) and returned it to France for reprocessing as well as conditioning the low and intermediate level radioactive waste (such as silo graphite).

The ministerial order also requested that the Spanish national radioactive waste company (Empresa Nacional de Residuos Radiactivos, S.A. — ENRESA) draw up a decommissioning and dismantling plan and that this be submitted to the Ministry of Industry and Energy for approval. The order also stipulated that when the plan had been approved by the Ministry and following a favourable report by the Nuclear Safety Council, ENRESA would be empowered to become the operating organization responsible for the plant in order for it to carry out the dismantling tasks.

In June 1991, ENRESA submitted a document analysing the different alternatives for decommissioning of the plant, along with the most feasible proposals, taking into account the different aspects of conditioning and dismantling. In 1992, following a favourable report by the Nuclear Safety Council, the Directorate General for Energy accepted the dismantling alternative proposed by ENRESA which basically consisted of dismantling all the plant structures, equipment and systems except for the nuclear reactor shield and other protective elements.

At the same time the Ministry of Industry and Energy requested that ENRESA submit a decommissioning and dismantling plan within 18 months. In May 1994 ENRESA submitted the Vandellós 1 dismantling plan to the Provincial Directorate of the Ministry of Industry and Energy in Tarragona. Following approval of this plan, a favourable report by the Nuclear Safety Council and approval of an environmental impact statement by the Ministry of the Environment, ENRESA will become the operating organization responsible for the NPP until completion of the work.

Subsequently Hifrensa, as plant owner, will take charge of the free release part of the site following a check by the authorities, the Nuclear Safety

Council and the Ministry of Industry and Energy that the work performed by ENRESA is in compliance with the dismantling plan. The Vandellós 1 NPP is the first plant to be dismantled in Spain and one of the first commercial power plants to be dismantled anywhere.

I–10.2. Decommissioning strategy

Prior to submission of the plan, ENRESA carried out various preliminary studies on the status of the facility, the technologies available and the experience acquired from the decommissioning and dismantling activities at similar plants in other countries. Three main decommissioning strategies were studied:

(1) Indefinite maintenance in the shutdown state;
(2) Safe enclosure of the defuelled reactor vessel and its contents, together with decontamination of most of the rest of the site;
(3) Immediate dismantling and release of the whole site.

The Ministry of Industry and Energy chose the second option and stipulated that once the spent fuel and radioactive operating wastes had been removed, most of the structures and components outside the reactor shield would be dismantled, except those which ensured confinement of the shield up to the end of the latency period. This option offered several advantages over the other alternatives, including reductions in risk and potential failures of systems and components. It minimized both the generation of secondary wastes and the costs of implementation, surveillance and conservation. The reactor shield will be suitably isolated and all external connections and penetrations will be closed, blocked and sealed allowing the residual activity contained in them to decay with time.

A series of activities will be undertaken to keep the shield safe, which will include carrying out structural conservation/degradation studies, installing suitable surveillance and control systems and implementing the maintenance plan to guarantee safe isolation of the reactor during the latency period. The site surveillance boundary will be reduced by about 80% by following this decommissioning and dismantling strategy. From the technical and operational points of view the dismantling project is considered to be very feasible since it is to use straightforward technologies and well proven working methods and tools. The different phases of the dismantling project are shown in Table I–VI.

TABLE I–VI. DISMANTLING SCHEDULE

Phases	1998	1999	2000	2001	2002
Preparatory activities	■				
Dismantling of active parts		■	■	■	
Decontamination of major items and zones		■	■	■	
Isolation of the reactor		■	■		
Scarifying of wall surfaces		■	■	■	
Dismantling of graphite silos			■	■	■
Disassembly of conventional items		■	■	■	■
Reactor building and latency period infrastructures			■	■	■
Dismantling of other buildings and installations		■	■	■	■

I–10.3. The post-operational phase (transition phase)

The post-operational transition phase includes the periods during which Hifrensa proposed keeping the plant in a safe shutdown condition and the period addressed by ENRESA prior to the actual dismantling, called 'preparatory activities' in Table I–VI. The preliminary activities carried out by Hifrensa included the following:

(a) Defuelling of the reactor and removal of fuel from the site.
(b) Conditioning of operational wastes and extraction and preconditioning of wastes kept in the graphite silos.
(c) Draining the various circuits; in some cases this was followed by a decontamination of the circuits in order to reduce the dose rates for subsequent dismantling work.
(d) Conventional dismantling operations, such as those affecting the CO_2 tanks, steam generator and the main turbine (Fig. I–28).

FIG. I–28. Vandellós NPP unit 1, cutting steam generator pipes.

(e) Taking non-essential systems out of service.
(f) Reduction or removal of the fire hazard and risk of possible uncontrolled release of toxic substances.
(g) Radiological characterization to provide radiological information on the shut down reactor, to facilitate decisions on other decommissioning steps such as decontamination, dismantling and removal of components and equipment, demolition of structures, management of decommissioning waste, estimates of future radionuclide inventories and the funding of decommissioning activities.

The preparatory activities carried out by ENRESA in 1998 included installation of new systems and facilities, modifications to existing systems and removal of specific conventional systems and components in order to meet the requirements of the decommissioning and dismantling plan. These nuclear and non-nuclear activities included:

(1) Removal of systems not required for dismantling of the reactor;
(2) Removal of inflammable and/or toxic components and systems;

(3) Modification of the electrical systems and installation of a new electricity distribution system to meet future dismantling needs;
(4) Installation of new ventilation systems in radiologically controlled areas;
(5) Replacement of the control room with an integrated surveillance post;
(6) Transfer of the management and administration offices to the eastern end of the site in order to clearly separate administrative and works related activities;
(7) Modification of the effluent treatment and dilution system;
(8) Modification of the materials decontamination workshop;
(9) Modification of the fire fighting system and training of the fire brigade;
(10) Construction of the materials cutting and conditioning workshop;
(11) Design and construction of a storage building to lodge the radioactive wastes generated during the dismantling;
(12) Enhancement of the medical service and its equipment;
(13) Installation of a new entrance to the reactor building and a radiological control post at the exit;
(14) Installation of a weighing station and radiological control gate monitor for transport vehicles leaving the site;
(15) Amendment of the management/operating organization to meet future decommissioning and dismantling requirements as shown in Fig. I–29; Equipping of the materials management cutting workshop;

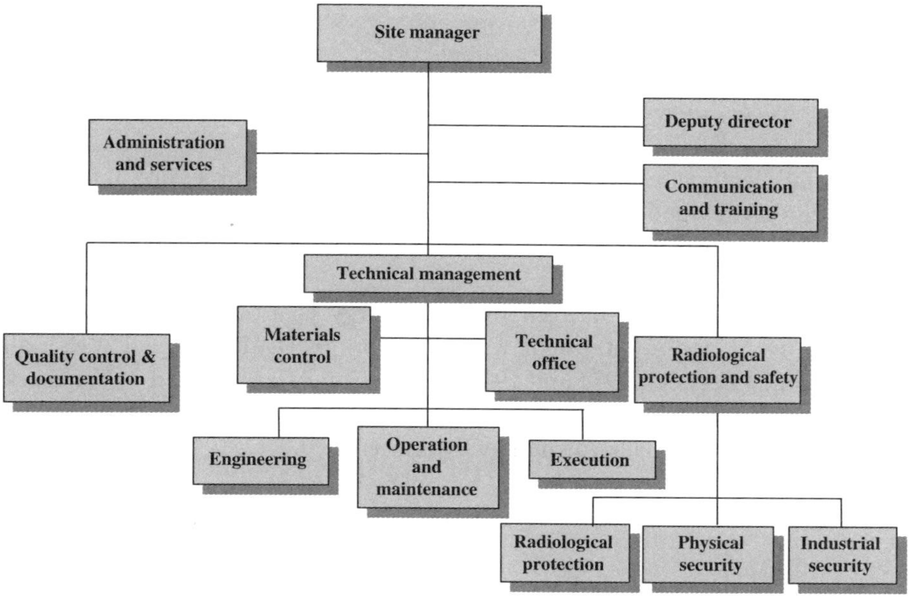

FIG. I–29. Organization during the dismantling of unit 1 of the Vandellós NPP.

(16) Dismantling and demolition of the plant's auxiliary electrical systems building;
(17) Removal of fuel tanks and demineralized water tanks;
(18) Disassembly of neutralization and chlorination pits and other minor exterior installations;
(19) Demolition of the discharge channel, the demineralized water tank foundations and the fuel transfer canopies.

I–10.4. Conclusions

The transition period at the Vandellós NPP has been essential to the overall development of the decommissioning project in that it has allowed infrastructures and systems to be adapted from their original operational status to meet the future needs of decommissioning and dismantling. In particular, it has been used to study, review and modify the organizational management arrangements, the documents needed for future plant operation, site characterization aspects, nuclear and non-nuclear plant systems and components that are both required or affected by the decommissioning and dismantling programme.

I–11. THE TRANSITION PERIOD AT UNIT 1 OF THE BARSEBÄCK NPP, SWEDEN

I–11.1. Introduction

I–11.1.1. Nuclear power in Sweden

Total electrical energy production in Sweden is in excess of 150 TW·h, almost equally divided between nuclear and hydroelectric power. Sweden also imports and exports electricity as required, principally due to the ready availability of hydroelectric power.

The total net output from nuclear energy is around 10 000 MW(e), provided by 11 operational light water reactors (8 BWRs and 3 PWRs), all commissioned between 1972 and 1985. These reactors are located at Forsmark (owned by Vattenfall), Oskarshamn (owned by OKG AB), Ringhals (owned by Vattenfall) and Barsebäck (owned by Sydkraft) [I–7].

I–11.1.2. Decommissioning of Barsebäck 1

In 1997 the Swedish Parliament passed legislation on the decommissioning of nuclear energy. The legislation gives Parliament the right to decide which nuclear reactors shall cease operations. On 5 February 1998, under this legislation, the Swedish Government took the decision to close Barsebäck from June 1998. An appeal to the Supreme Administrative Court resulted in a temporary postponement of the closure. The Court subsequently declared that the Government's decision should stand and Barsebäck 1 was closed on 30 November 1999.

The management company at Barsebäck 1, together with the trade unions, initiated a series of projects called Framtidsfabriken/Nya Fabriken (the factory of the future/the new factory) whose purpose was to conduct a broad pilot study of possible future scenarios and to consider these from the perspective of the company and the individual. The work was done with extensive participation of the staff at Barsebäck Kraft AB, resulting in:

(a) Proposals for changes to the company's quality system,
(b) Proposals for future changes to the organization and appropriate management arrangements,
(c) Proposals for meeting future skill requirements,
(d) Establishment of a mentoring system to pass on skills and experience,
(e) Ensurance of security for employees and safeguarding resources for Barsebäck Kraft AB through different forms of employment guarantees,
(f) Opportunities for agreed retirement schemes for those over the age of 60.

The possible future closure of Barsebäck 2 depends upon the balance between the resultant loss of electricity being compensated for by new electricity generation and reduced electricity consumption. This balance could be achieved through measures such as increasing the supply of electricity from renewable energy sources and through reduced electricity consumption. The Government made an assessment in 2001 which found that closure could not be fully justified without further review of its implications concerning future shortfalls in national energy requirements, electricity prices, supply of electricity to industry, the power balance, the environment and the climate. A further review was planned [I–8 to I–11].

I–11.2. Barsebäck

I–11.2.1. The plant

The Barsebäck NPP, owned by Sydkraft AB, is located in southern Sweden along the coast opposite Denmark. It comprises the following reactor units:

(a) *Barsebäck 1*
Type: BWR,
Capacity: 615 MW(e),
Start of operation: 1975,
Supplier: ABB Atom (currently Westinghouse Atom AB),
Owner: Sydkraft,
Operating organization: Barsebäck Kraft AB, Vattenfall,
Production: Total 100 TW·h (1999),
Status: Permanently shut down since 30 November 1999, decommissioning is ongoing.

(b) *Barsebäck 2*
Type: BWR,
Capacity: 615 MW(e),
Start of operation: 1977,
Supplier: ABB Atom (currently Westinghouse Atom AB),
Owner: Sydkraft AB,
Operating organization: Barsebäck Kraft AB, Vattenfall,
Production: Total 99 TW·h (2001),
Status: Fully operational.

Barsebäck 1 and 2 are adjacent to one another and are connected by buildings that house electrical systems, processing systems, control rooms and staff rooms. For reasons of reactor safety it has been decided that demolition of Barsebäck 1 will not begin as long as Barsebäck 2 is operational.

I–11.2.2. Special arrangements

Several complex agreements were reached between the Swedish Parliament and Sydkraft, the owners, before decommissioning of Barsebäck 1. In broad terms these involved:

(a) Sydkraft AB remaining as the owner of the Barsebäck 1 and 2 plants and being responsible for future demolition, spent fuel and demolition waste;
(b) Barsebäck Kraft AB, previously a subsidiary of Sydkraft, now being part of the operations of the government owned electricity generator Ringhals Vattenfall AB;
(c) 74% of Ringhals/Barsebäck Kraft AB to be owned by the government owned Vattenfall, with 26% owned by the privately owned Sydkraft;
(d) Barsebäck Kraft AB being responsible for decommissioning and service operations, as well as producing plans for the future demolition of Barsebäck 1;

(e) The Swedish nuclear authorities (the National Swedish Nuclear Power Inspectorate and the Swedish Radiation Protection Institute) deciding that the Barsebäck nuclear operating licences should be held by Barsebäck Kraft AB, with all official contact with the various authorities being the responsibility of these parties [I–12, I–13].

I–11.2.3. The organizational structure of Ringhalsgruppen

On 1 April 2002 Ringhals and Barsebäck Kraft AB merged to form a new organization called Ringhalsgruppen which has a total of 1500 employees, 1375 from Ringhals and 125 from Barsebäck Kraft AB. While the two companies remain separate legal entities, Ringhalsgruppen has a single Chief Executive Officer and comprises five separate production units, R1, R2, R3, R4 plus an additional B2 unit, together with one decommissioning unit, B1, as shown in Fig. I–30. Barsebäck Kraft AB (BKAB) has its own independent safety and environmental staff and operates a production unit (B2) which has overall responsibility for Barsebäck 2 operations, and a decommissioning unit (B1) that has 6 employees and has overall responsibility for the shut down Barsebäck 1 plant. Services such as maintenance, projects and administration are procured from the implementation side of Ringhals AB.

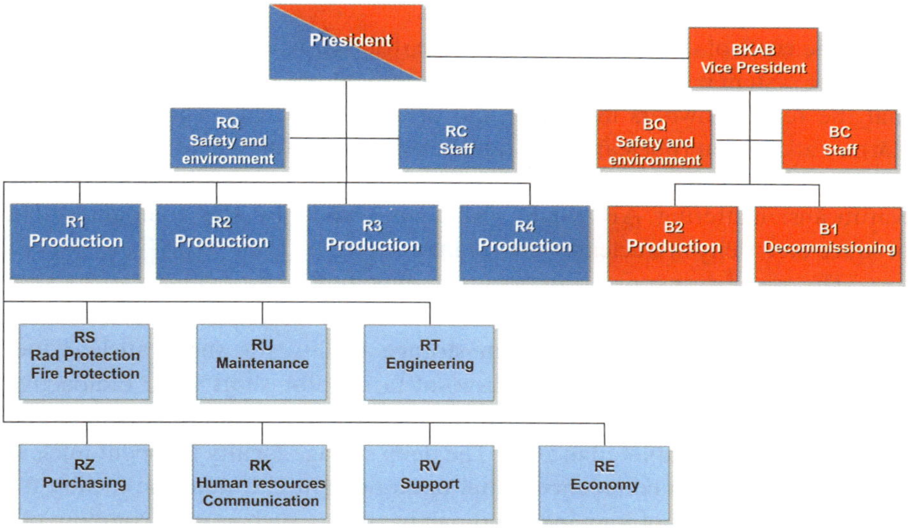

FIG. I–30. Organization of Ringhalsgruppen.

I–11.3. Legislation and authorities

I–11.3.1. Legislation

I–11.3.1.1. Kärntekniklagen (1984:3) (Nuclear Technology Act)

This legislation applies to nuclear energy activities such as the construction, ownership or operation of NPPs. According to this legislation a company with permission to operate a nuclear power plant is fully responsible for the secure handling and ultimate disposal of nuclear waste. This also includes future demolition of the nuclear power plant as well as conducting the necessary research and development work associated with ultimate disposal methods. According to the framework agreement reached between the Swedish Parliament, Sydkraft AB and Vattenfall AB the responsibility for Barsebäck lies with its owners, Sydkraftkoncernen [I–14].

I–11.3.1.2. Strålskyddslagen (1988:220) (Radiation Protection Act)

The purpose of this act is to protect people, animals and the environment from the damaging effects of radiation, including both ionizing and non-ionizing radiation [I–15].

I–11.3.1.3. Finansieringslagen (1992:1537) (financing of future costs for spent nuclear fuel, etc.)

The costs of demolition and ultimate disposal will be handled through a financing system based on a fee to the government per kilowatt-hour produced that the reactor owners include in the price of electricity. The Government Kärnavfallsfonden (nuclear waste fund) manages the funds that are built up in this way. The disposal of all nuclear waste is the responsibility of Svensk Kärnbränslehantering AB (SKB), which is owned by the companies that produce electricity at nuclear power plants.

The waste transportation arrangements and facilities include the ship M/S Sigyn, a final disposal facility for short lived operational radioactive waste (SFR) at Forsmark and an interim storage facility for spent nuclear fuel (CLAB) at Oskarshamn. A final disposal facility for short lived radioactive dismantling and demolition waste (SFR-3) has yet to be built at Forsmark and is expected to be completed in 2015. The deep storage facility for spent nuclear fuel is anticipated to commence initial operations in 2015 and the facility for the ultimate disposal of long lived radioactive waste, such as core components, is expected to be ready in 2045. To date the nuclear waste funds have enabled

SKB to build and operate CLAB, the M/S Sigyn transport system, the Äspö rock laboratory and the associated costs of research and development [I–16 to I–18].

I–11.3.1.4. Miljöbalken (The Environmental Code)

The Environmental Code came into force on 1 January 1999 and provides co-ordinated, broadened and enhanced environmental legislation for sustainable development. It consolidates the regulations from fifteen previous environmental laws [I–19].

I–11.3.2. Authorities

I–11.3.2.1. Swedish Nuclear Power Inspectorate (SKI)

The Swedish Nuclear Power Inspectorate is the national authority that checks to ensure that the nuclear licence holders meet the necessary plant operational safety requirements, including control of nuclear material plus the handling and ultimate disposal of nuclear wastes. The Swedish Nuclear Power Inspectorate also contributes to the development of nuclear safety work [I–20].

I–11.3.2.2. Swedish Radiation Protection Authority (SSI)

The Swedish Radiation Protection Authority is responsible for ensuring that the damaging effects of radiation on people and the environment are minimized [I–21].

I–11.3.2.3. National Swedish Environment Protection Board

The National Swedish Environment Protection Board is a government environmental authority that works towards ecologically sustainable development. The Swedish Government has commissioned the Board to co-ordinate and be the driving force in environmental work both nationally and internationally [I–22].

I–11.3.2.4. County Council

Each county in Sweden has a County Council, a government authority whose principal task is to act on behalf of the Parliament and to implement government decisions.

One of the roles of the Council is to act as the voice of its constituents and ensure that the county is developed according to the wishes of its population. This role includes monitoring the environment by inspecting industries, environmentally hazardous waste, health aspects and energy requirements. Each County Council has a delegation that scrutinizes the environment as required by the Environmental Code [I–23].

I–11.3.3. Licensing requirements [I–19, I–24]

I–11.3.3.1. Swedish Nuclear Power Inspectorate

Regulation SKIFS 1998:1 requires that the measures necessary to contain nuclear waste be documented in a plan before the decommissioning and demolition of a plant commences. This plan has to be safety checked, assessed and approved by SKI.

I–11.3.3.2. Swedish Radiation Protection Authority

The SSI is responsible for preparing the necessary planning regulations prior to and during the decommissioning of an NPP.

I–11.3.3.3. County Council and National Swedish Environment Protection Board

An environmental impact description (miljökonsekvens beskrivning (MKB)) has to be produced by the power plant owner before the commencement of demolition. It has to highlight a broad spectrum of issues, including describing the direct and indirect effects of the demolition activities on people, animals, plants, the ground, water, air, climate and the countryside.

I–11.3.4. Licence approval

Currently, under the Nuclear Technology Act, no further licence is required to decommission and demolish a Swedish NPP as these activities are already covered by the operating licence. Hence, by law, this provides the owners with the right to decommission and demolish a plant within the framework of the licence already issued without resort to further licences. However, future decommissioning and demolition will almost certainly be taken forward following an assessment of the decommissioning and demolition methods and their environmental consequences.

I–11.4. Barsebäck 1 decommissioning strategy

A strategic plan for the decommissioning of Barsebäck 1 has been produced and submitted to SKI and SSI.

The plant owner, Sydkraft AB, is responsible for the future demolition of Barsebäck 1 and 2. Figure I–31 details the strategic planning for decommissioning the Barsebäck NPP. Demolition of Barsebäck 1 will probably not begin before 2020 due to the following factors:

(a) Demolition of Barsebäck 1 will not commence as long as Barsebäck 2 is in operation. On the basis of a 40 year service life, Barsebäck 2 will not be shut down before 2017.
(b) Barsebäck 1 and 2 will probably be demolished as a joint project and thus the start of demolition could be further postponed.
(c) The SFR-3 facility for the final disposal of short lived radioactive demolition waste is not expected to be completed before 2015.

In light of these factors it is not considered appropriate at present to commence comprehensive or detailed planning of the actual demolition phase for B1. In the immediate future it is more important to carry out a number of initial investigations and to produce the plant documentation that will be needed for procuring and implementing the demolition work and, in parallel, develop general skills required for the demolition process. Pre-project activities such as project structuring, environmental impact descriptions, production of safety reports, gaining approval, etc. for the final demolition phase are planned to begin 4 to 5 years before the implementation phase [I–25].

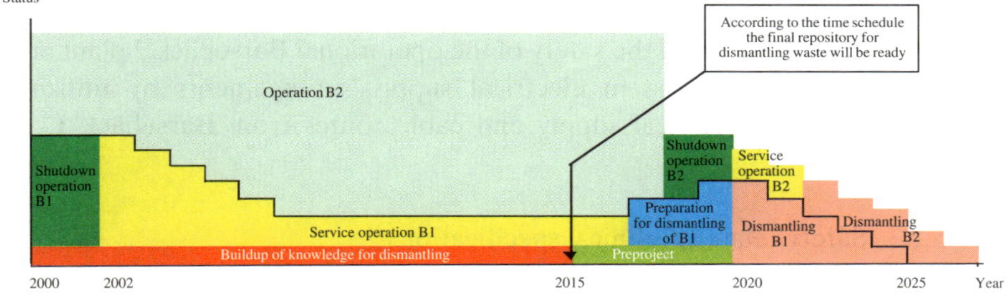

FIG. I–31. Planning scenario for decommissioning of the Barsebäck NPP.

I–11.5. Barsebäck 1 shutdown/service operations

I–11.5.1. General

Barsebäck Kraft AB was reorganized in May 2000 and produced new organizational and management criteria for conducting decommissioning activities in parallel with ongoing operations. The new integrated organization, outlined above, was formed on 1 April 2002.

Following the removal of all the fuel from the reactor in late 2001, Barsebäck 1 has been in a so called 'service operations' period and will remain in this phase for a considerable period. The main aim of this phase is to prevent the inadvertent spread of radioactivity into the surrounding areas and to minimize the impact on the continued operation of the Barsebäck 2 reactor [I–25 to I–29].

I–11.5.2. Completed activities

I–11.5.2.1. Fuel aspects

All the fuel was discharged from the reactor immediately after shutdown. A new project was also initiated to reuse the fuel elements and transport these to Barsebäck 2 with the remaining spent fuel being transported to CLAB. The fuel was transferred between Barsebäck 1 and 2 using new techniques which involved transferring the fuel in fuel containers that were not completely filled with water, thereby allowing a certain degree of expansion caused by potential pressure and temperature increases. This method of fuel transfer was very successful and the resultant pressure and temperature increases were minimal. These activities went entirely according to plan and were completed in December 2001, the start of the service operations period for Barsebäck 1.

I–11.5.2.2. Project 327 EB 1

This project enhanced the safety of the operational Barsebäck 2 plant and its auxiliary feedwater system electrical supply arrangements by utilizing redundant systems, electrical supply and cable routes from Barsebäck 1, as required.

I–11.5.2.3. Safety related technical specifications

Safety related technical specifications provide the approved framework within which allowed reactor plant operations take place with due regard to the safety of the surroundings and the environment. The likelihood of a serious nuclear accident at the Barsebäck 1 plant has been effectively reduced to zero as all nuclear fuel was removed by 12 December 2001.

However, as there are still residual radioactive products at the plant the specifications concentrate on preventing uncontrolled emissions of radioactive material or airborne radioactivity into the surroundings and the environment, radiological protection of plant employees and ensuring that Barsebäck 1 does not jeopardize safety at the operational Barsebäck 2 plant.

I–11.5.3. Current activities

I–11.5.3.1. Monitoring and maintenance during the service operations period

Some systems and components need to be kept operating during the entire service operations period. An evaluation is currently in progress to determine the number of components and spare parts needed for both Barsebäck 1 and 2. For Barsebäck 2 the main criteria are operational availability, audit requirements, unplanned downtime, new acquisition costs and component obsolescence, as well as the capability of the various manufacturers to deliver components. The aim of this evaluation is to identify, secure and preserve all vital strategic components and spare parts.

The required status of the plant just prior to planned demolition is also being examined to determine the amount of plant maintenance needed during the service operation period. The actual maintenance that will be carried out will depend on several factors such as safety and costs. This examination will determine whether it is safe, feasible and cost effective to maintain certain systems at current or reduced levels or to do nothing and eventually reinstate them to the required status just prior to demolition.

For example, the main power generators and turbines have been preserved as they have a large number of electric motors and pumps, and it has been determined that the transformers can be shut down for a long period of time without having to take special measures. Since plant closure in December 1999, the diesel generators and electrical apparatus room have been maintained as operational units and have been used to provide an alternate and separate supply to Barsebäck 2's auxiliary feedwater system following RA2 (shutdown for overhaul at Barsebäck 2) in 2001.

Some monitoring will also be required in the future. This will be carried out by the Barsebäck 2 operational shift staff as there will no longer be any permanent monitoring staff in the control room at Barsebäck 1. All necessary information will be routed to and monitored in the Barsebäck 2 control room.

Personnel will also make routine rounds in Barsebäck 1 during the normal working day.

I–11.5.3.2. Profiling of radiological activity

Radiological activity will be profiled at the Barsebäck 1 plant to provide data for assessing future requirements for decontamination and waste categorization.

I–11.5.3.3. Waste inventory

Following contamination and radiological activity profiling, a plant specific inventory of waste quantities will be produced for Barsebäck 1 by about 2005. This will be an update of an earlier SKB general summary of typical waste volumes and quantities associated with demolition of a nuclear power plant.

I–11.5.3.4. Decommissioning documentation

Previous national and international experience indicates that prior to the demolition phase there has tended to be a shortfall in related plant documentation. Work is ongoing to clarify what documentation is required and ensure that it will be available as required to meet the planned decommissioning, dismantling and demolition project timescales. This work is scheduled to be completed by the end of 2003.

I–11.5.3.5. The CONMOD project (reactor containment matters)

The European Commission and a four party consortium including Force Technology (Denmark), Scanscot Technology (Sweden), Electricité de France (France) and Barsebäck Kraft (Sweden) have agreed to carry out a joint project called Concrete Containment Management using the Finite Element Technique Combined with In Situ Non-destructive Testing of Conformity with Respect to Design and Construction Quality (CONMOD) to look into reactor containment matters. It will be carried out within the framework of the Euratom programme which covers research and training in the field of nuclear energy.

The project commenced on 1 January 2002 and will require three years. It involves investigations into the safety aspects of concrete structures such as containments at NPPs, which are subject to ageing processes that can reduce their safety as well as their functional lifetimes. Serious problems with these structures have been known to result from defects caused at the construction stage. In order to learn the actual status of concrete containments it is necessary to apply investigative techniques that are capable of providing information about the internal structure and condition of the concrete and its reinforcement.

I–11.5.3.6. Developing demolition skills

The purpose of this study is to benefit from national and international experience in reactor demolition techniques by creating a contact network of authorities, power plants, experts, firms and others. This study has no set time frame as it intends to retain current and future state of the art technologies, methodologies and techniques right up to the commencement of actual demolition.

Its purpose is to build up the knowledge base so that BKAB can deploy the most up to date skills and resources to produce the necessary decommissioning documentation, secure the demolition licence and procure, control and monitor the consultancy firm and contractors who will carry out final demolition. Activities carried out to date include:

(a) Forming a national demolition group under SKB's management and comprising representatives from the Swedish nuclear power companies. The purpose of the group is to provide best estimates of future costs for decommissioning power plants by focusing on demolition technologies, logistical issues, the best technical solutions, and the processing and handling of radiological wastes.
(b) Making contact with other nuclear power plants worldwide currently undergoing some form of decommissioning such as the NPPs at Ignalina (Lithuania), Greifswald and Würgassen (Germany), Risø (Denmark) and Vandellós 1 (Spain).
(c) Contacting various contractors in order to understand their current and future demolition methods.
(d) Carrying out various decommissioning related assignments for the IAEA in Vienna and the World Nuclear Association in London.

I–11.5.3.7. Future regeneration of the site

The Framtidsfabriken/Nya Fabriken project outlined above has investigated potential new business uses for the redundant Barsebäck 1 reactor. Subject to further studies, these could include its future use as a spare parts depot, a service and training centre, an experimental plant or facility for testing new technologies, school related activities, or for investigating pilot decommissioning projects.

I–11.5.3.8. Personnel matters

The closure of Barsebäck 1 and the uncertainty concerning BKAB's future has weighed heavily on many of its employees. The most important challenge has been to maintain employee motivation and professionalism despite this uncertainty to ensure that the future safety of the operational Barsebäck 2 reactor is not compromised.

Changes have been made to the organizational and management structure; employment guarantees have provided security; an employee mentoring system has been set up and a design school has been established giving employees the opportunity to be retrained for other duties. These and other measures have resulted in an open and positive dialogue between management and staff that has given individuals a greater sense of security and choice when faced with an uncertain future.

I–11.6. Conclusions

The personnel, technical, logistical and management initiatives taken to date will enable the operating company to maintain the necessary skills for continued reactor operations at Barsebäck 2 while beginning the decommissioning project at the adjacent Barsebäck 1 reactor. Its position has been further strengthened by Ringhals' and Barsebäck's new robust management and organizational arrangements and has been greatly assisted by the ongoing dialogue between the authorities, which have expressed a positive attitude to new decommissioning work under way at Barsebäck 1.

A number of specific initiatives have either been carried out or are under way to improve future decommissioning requirements and maintain and improve existing nuclear safety standards at the operating reactor. These cover personnel skills, training, plant modifications, maintenance requirements, waste characterization studies, monitoring arrangements and containment studies. Important studies are also under way to learn the best national and international practices for future decommissioning, dismantling and demolition, together with ongoing investigations into potential new uses for the decommissioned site.

I–12. TRANSITION FROM OPERATION TO DECOMMISSIONING OF THE JASON REACTOR, UNITED KINGDOM

I–12.1. Introduction

Jason was a water/graphite moderated low power training and research reactor of the Argonaut type, used at thermal power levels of up to 10 kW. It was first taken critical at the Royal Naval College, Greenwich, London, in November 1962, having been previously operated by the Hawker Siddley Nuclear Power Corporation at Langley in the UK from February 1959. It was used for training Ministry of Defence Naval (MoD(N)) personnel and others involved in the naval nuclear submarine propulsion programme and for research as a source of neutrons for irradiation and shielding investigations. The decision to decommission Jason to the (then) IAEA stage 3 status (unrestricted site use) was taken by the MoD(N) in 1996 following the ministerial decision that the whole college would pass to non-defence use by the millennium.

Jason was situated in King William building, which is a grade 1 listed building within the college, which itself is a scheduled ancient monument having world heritage site status. It had been operated at the college by the MoD(N)'s Department of Nuclear Science and Technology (now called the Nuclear Department, HMS Sultan) and was last shut down for annual maintenance in July 1996. Prior to the decision to decommission, all fuel modules had been removed and stored in the adjacent fuel storage pits within the reactor hall, and extensive mechanical and electrical modifications and remedial work were under way on the reactor hall crane. The decision to decommission Jason was taken prior to completion of work on the crane.

Following the decision to decommission, a MoD(N) headquarters project sponsor was appointed and three main civilian contracting firms were selected through open competition to carry out project management, defuelling and reactor dismantling activities. The civilian Jason project manager, from AEA Technology (now Serco Assurance), was appointed in October 1996 with a remit to produce an early viability report and decommissioning plan and to project manage the overall decommissioning programme to (then) IAEA stage 3 status. An AEA Technology project manager (fuel removal) was appointed to manage the removal of the Jason reactor fuel modules and their subsequent transfer to British Nuclear Fuel Ltd's (BNFL) Sellafield facility. NNC Holdings Limited (NNC) were appointed as overall principle or prime contractor and project manager (reactor dismantling) with a remit that included reactor dismantling, waste removal and subsequent radiological cleanup of the site [I–30, I–31].

All such contractual appointments took place during the transition period (Fig. I–32).

Following completion of POCO in June 1998, extensive fuel removal and supporting equipment was erected and commissioned on-site and all fuel was subsequently loaded into a UKAEA owned Unifetch fuel transfer flask within the reactor hall. The flask was then removed from the building and subsequently transported by road to BNFL Sellafield in September 1998. The completion of POCO and fuel removal marked the end of the transition phase and the beginning of the dismantling phase.

The prime objective of the project was to safely decommission Jason to time, cost and required quality [I–32, I–33]. The appointment of three main but separate decommissioning contractors dictated the requirement for a suitable and workable safety management system that satisfied the site licensee's mandated nuclear and radiological safety responsibilities. Many new decommissioning specific safety documents were written and the Jason operational safety management arrangements and documents were extensively modified to meet the new decommissioning requirements. These new arrangements ensured that the site licensee's nuclear and radiological safety responsibilities were not compromised during each stage of the decommissioning process. They also ensured that due process occurred for the approval of all safety related documentation and that there was always adequate control of physical work on the ground.

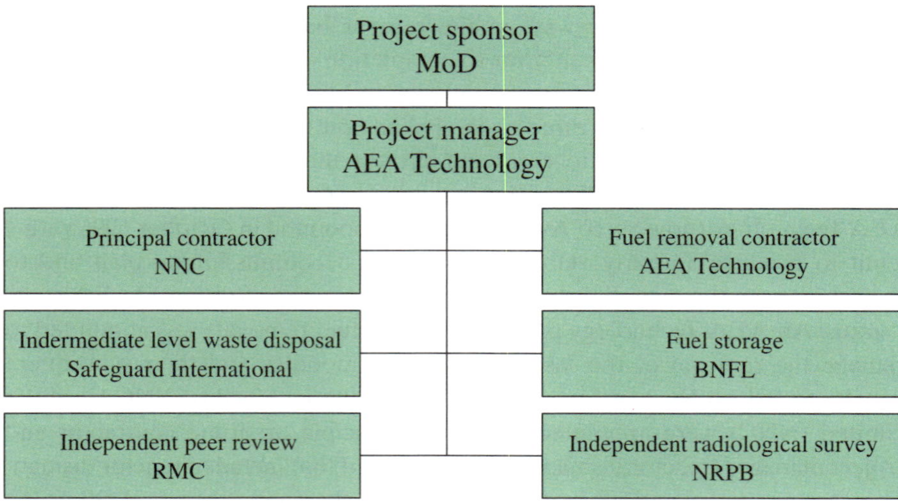

FIG. I–32. The decommissioning project management team.

I–12.2. Timescales

The three year deadline for completing the Jason decommissioning project and subsequently vacating the college by the millennium was very short in traditional reactor decommissioning terms. The management of the decommissioning project was undertaken at the same time as the management of another major project to relocate the Department and its training facilities from Greenwich, London to HMS Sultan, Hampshire. In addition, continuous student training was required to be maintained throughout, initially at Greenwich and, following relocation, at HMS Sultan. This had the potential to reduce nuclear safety margins unless significant safety management resources and arrangements were put in place.

The site licensee set up a series of dedicated safety management working groups and decommissioning procedure authorization groups to augment the normal operational management arrangements. The headquarters sponsoring organizations also set up external project, press, public relations and local planning approval teams. Additional nuclear engineering and safety personnel were also recruited by the site licensee to augment the existing reactor operational staff, whose remit was to concentrate solely on the decommissioning project.

I–12.3. Decommissioning plan

From the outset of the Jason decommissioning project is was clear that an overall decommissioning plan would be required that interpreted and referenced MoD(N), international and national nuclear safety policies and provided the overall planning framework and safety justification for the tasks undertaken during decommissioning. The first issue of the Jason decommissioning plan was approved in June 1997. Subsequent to that, a further three issues were produced as the overall decommissioning project progressed. The plan was just one of a whole set of safety documents that were produced as part of the preparations for decommissioning.

The Ministry of Defence (MoD) had no previous practical experience in decommissioning a shore based research reactor to unrestricted release status or in producing and executing the necessary decommissioning plan. The contracted AEA Technology Jason project manager had extensive previous experience of reactor decommissioning projects and one of his first tasks on appointment was to produce a viable decommissioning plan.

The plan provided the decommissioning framework by giving an overview of the whole project, including the selected strategy, a description of the decommissioning tasks and the means whereby they would be managed in a safe, timely and efficient manner. It also incorporated the decommissioning programme, which provided a clear statement of the decommissioning tasks and their schedule. The individual sections of the plan consisted of the executive summary, introduction, description of the Jason facility, radioactive inventory, hazardous substance inventory, as fitted equipment and systems list, IAEA and national decommissioning strategies, the selected decommissioning strategy, objectives, operations, methods, procedures, waste, environmental monitoring, surveillance, safety management, safety documentation, resources needed and the overall decommissioning programme.

The decommissioning plan was fundamental to carrying out a safe, timely and efficient decommissioning project in that it provided the framework for the whole project, including the selected strategy, the decommissioning tasks, the means of executing those tasks and the necessary safety management arrangements.

I–12.4. Operational safety case

The existing operational safety case for Jason consisted of a single overarching safety document originally written in the early 1960s. This document had been extensively revised over the years but retained its original format and was supported by tiers of lower level documentation. The last major revision of the safety document occurred in 1994 when it was brought up to modern safety report standards. The document consisted of 19 sections covering all aspects of the layout, structure, systems, power supplies, facilities, waste management, radiological protection, safety management, safety principles and safety criteria, hazard analysis, deterministic and probabilistic safety assessments, operation, commissioning, modification, quality assurance and decommissioning. Each section referred to other safety management or controlling documents as necessary.

The decommissioning section of the safety document consisted of a basic plan derived from the IAEA and Nuclear Installations Inspectorate (NII) guidelines available at that time, covering outline regulatory requirements, waste disposal, transport, environmental assessments, decommissioning plans, safety evaluation, administrative controls, quality, emergency and security arrangements. A new decommissioning safety case was worked up very early on in the project, under contract by the AEA Technology Jason project manager. It included consideration of the latest international and national decommissioning guidelines and industry experience gained from previous low power and research reactor decommissioning activities. This process took a considerable amount of time and resources, which could have been significantly reduced had the existing operational safety case been more comprehensive and kept up to date with future decommissioning in mind.

I–12.5. Decommissioning safety case

The new Jason decommissioning safety case somewhat mirrored that used for a standard nuclear construction project as, in common with most decommissioning projects, facilities and structures had to be built for decommissioning. While POCO was carried out under the operational safety case, using the standard modification procedures, the safety cases for the rest of the transition period and subsequent dismantling consisted of two separate but complementary sets of safety reports covering fuel removal and reactor dismantling activities which were physically carried out by two different contractors, AEA Technology and NNC. The overall decommissioning safety case is shown in Fig. I–33.

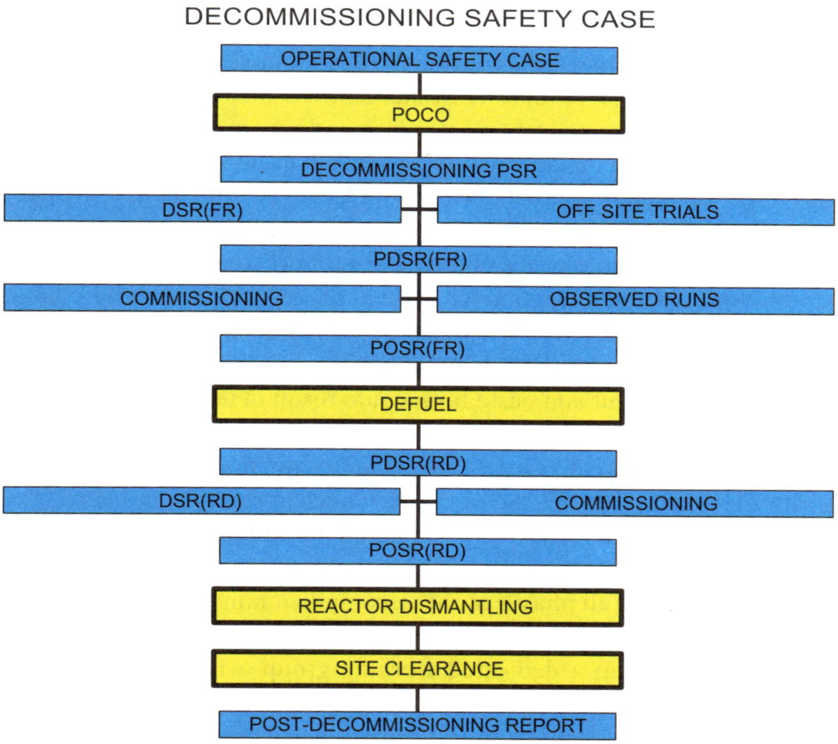

FIG. I–33. The Jason decommissioning project (SR = safety report, FR = fuel removal, RD = reactor dismantling, PD = pre-decommissioning, PO = pre-operational).

The safety case consisted of a preliminary safety report, a series of design substantiation reports covering both the defuelling and dismantling activities, two pre-decommissioning safety reports, a series of commissioning schedules and commissioning reports, two pre-operational safety reports and a single post-decommissioning report. The top level design substantiation reports consisted of the design report (fuel removal) and the design report (reactor dismantling).

Following POCO, the following reports sequentially constituted the Jason safety case:

(a) Fuel removal. The pre-decommissioning safety report (fuel removal) (PDSR), approved on 24 June 1998, and the pre-operational safety report (fuel removal) (POSR(FR)), approved on 11 September 1998.
(b) Reactor dismantling and waste removal. The pre-decommissioning safety report (reactor dismantling), approved on 17 November 1998 and the pre-operational safety report (reactor dismantling), approved on 17 December 1998.

The tight project timescales during the decommissioning of Jason dictated parallel production, review and assessment of the majority of safety reports associated with fuel removal and reactor dismantling, some of which took place during POCO. Different contractors produced each set of reports so the potential for document conflict was ever present. As an example of parallel assessment, the later issues of the POSR(FR) would be reviewed during the same period as the middle issues of the design report (reactor dismantling) and the early issues of the PDSR. Each individual report necessarily contained information from another report, which in itself was being amended and re-written by a different contractor. Inevitably, the potential for some information to be transferred in a cut and paste form could result in text from an early issue being carried over with mistakes that had been edited out of the current version of the document. Great care had to be taken by the reviewing authorities to ensure that this did not reduce nuclear safety margins.

This problem was effectively mitigated by appointing a single expert external independent peer review (IPR) team, under contract, to assess all documentation during all phases of the decommissioning project, regardless of its source of origin. The site licensee's internal review process was also enhanced by setting up a dedicated working group, which was effectively a subcommittee to the full nuclear site safety committee, whose remit was to review and assess all documentation prior to its submission to the full safety committee. The potential problem was further mitigated by the Jason project manager's review of all safety report related documentation prior to its submission to the site licensee's working group.

I–12.6. Client framework document

While the site licensee was authorised to operate Jason in accordance with the conditions of operation, the actual day to day safety management arrangements were detailed in mid to lower level documents. The plethora of orders, instructions and procedures contained within these documents had the potential to cause confusion, particularly among the decommissioning contractors.

A single non-executive safety management framework document was produced to provide the site licensee with a dedicated safety management focus describing how his mandated nuclear, radiological and conventional safety responsibilities would be managed during the decommissioning of Jason. This framework document outlined the key points from the various documents, covering departmental safety policy, regulatory control, quality assurance, safety management organization, key personnel, safety groups, committees, overall decommissioning policy, safety principles and criteria, facility area definitions, document peer review and approval, safety responsibilities, control of contractors, radiation protection, health and safety at work, audit and inspection, reporting and investigating incidents and accidents, and emergency response arrangements. Each section of the framework document referred to the applicable executive controlling document as required.

I–12.7. Control of contractors

In addition to AEA Technology and NNC, a further thirteen major and minor subcontractors were employed during the decommissioning of Jason, many of which carried out work during the transition phase. In order to maintain its mandated nuclear and radiological safety standards and responsibilities and to retain control of work on the ground, it was necessary for the site licensee to provide suitable and workable safety management arrangements and instructions to these various contractors. An organizational flowchart showing the principal and major subcontractors for the Jason project is given in Fig. I–34.

Contractor control instructions were written to ensure that the site licensee maintained its nuclear safety responsibilities. These instructions were produced in two parts. The first part specified the arrangements for controlling the safety of work by contractors within the Jason decommissioning facility and the second part specified how these arrangements would be complied with. The contractor control instructions encompassed training, qualifications and experience for safety related post holders, written systems of work, personnel protective equipment, safety monitoring, health and safety records, visitor safety, operating rules, maintenance, inspection, testing, categorization, document clearance procedures, occurrences, incidents, accidents and emergency arrangements. Other instructions addressed radioactive and non-radioactive waste management, movement of radioactive and dangerous materials, accounting for radioactive and nuclear materials, radiation, criticality and general safety, hazardous substances and facility security. In total, 21 individual contractor control instructions were produced.

I–12.8. Safety committees

All nuclear safety reports, design substantiation reports and most of the higher level nuclear safety management administrative documents had to be endorsed by the full Jason reactor safety committee and the naval regulatory authorities before being approved by the site licensee. The full safety committee had traditionally met on a regular basis about three times per year as dictated by the original Jason operating cycle. It was quite clear from the onset that the membership and frequency of meeting of the full safety committee had to be significantly increased to meet the tight project timescales. The site licensee considered that the frequency of committee meetings could be a potential treat to the timely completion of the programme, while recognizing

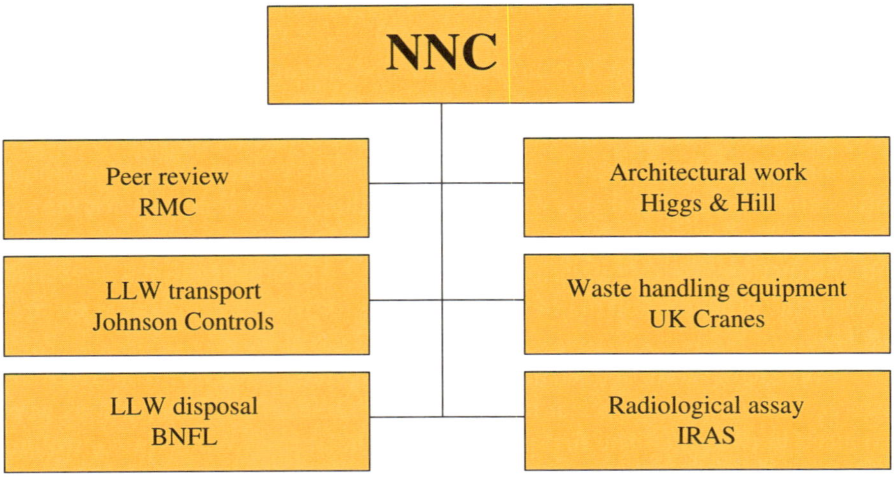

FIG. I–34. Principal and major subcontractors for the Jason decommissioning project.

that the committee's role was to provide advice and not necessarily to be driven by any time and cost considerations.

Following the decision to decommission, the full safety committee met six times in 1996, six times in 1997, seven times in 1998 and twice in 1999. Its external membership was increased from four to six to include two more members with previous experience of decommissioning activities. The site licensee chaired the committee, which consisted of the six standing external members, the site licensee's deputy and his Jason decommissioning superintendent (JDS). Up to four other senior staff members attended meetings, depending upon the agenda items.

The existing site licensee's safety document review arrangements were geared up to cater for the extensive internal and external peer review requirements, endorsement of safety related management documents and approval of formal modifications to the safety cases that would inevitably be required during the decommissioning project. To manage this due process effectively a site licensee staffed Jason safety advisory working group (JSAWG) was set up to provide the focus for these activities, effectively acting as the subcommittee to the full safety committee. Its chairperson was the site licensee's deputy and all members were either duly authorized persons (DAPs) or suitably qualified and experienced persons (SQEPs) from the site licensee's staff.

External IPR of all important safety related documents was undertaken prior to their submission to the JSAWG, carried out by an experienced consulting firm called RMC Consultants, under contract to the decommissioning project. This form of due process was very rigorous and often resulted in many document iterations. As every safety report and related document presented to the full site safety committee had to undergo prior external IPR and review by the JSAWG, the time spent obtaining full safety committee endorsement was considerably reduced and almost seamless. It was the rule rather than the exception that the respective contracted project manager and safety report author attended particular committee meetings to provide clarification and focus. In the event, the safety committee's considered advice to the site licensee was most rigorous and was instrumental in maintaining nuclear safety standards and the frequency of committee meetings met the tight project timescales.

I–12.9. Training

The site licensee had an extant nuclear training and requirements plan (NTRP) for Jason operations that listed all operational nuclear safety related posts and detailed what qualifications and training requirements were necessary for each post holder, the actual qualifications held and what training/

qualifications would be required in the future. Each post holder was designated as either a DAP or a SQEP, the appointment of which was approved in writing by the site licensee. The NTRP had to be significantly amended to cover decommissioning activities to include additional duties, extra client staff and, in particular, numerous civilian contractors. The civilian contractors did not have an equivalent to the NTRP.

The normal MoD contract placement process ensures that only suitably qualified contracting firms who meet particular quality standards, including individual employee training and experience requirements, will be appointed/contracted to carry out nuclear related work with the MoD. To meet the general requirements of the NTRP, AEA Technology and NNC were required to forward career histories and qualifications of employees related to each phase of the decommissioning project for consideration by the site licensee.

The site licensee formally approved the appointment of all Jason decommissioning contractors who were considered to be operating as DAPs, such as the individual project managers and radiation protection advisers. He 'noted' the appointment of all other contractor SQEPs such as all site operatives and off-site safety report authors. While the SQEP appointment was actually authorized by the chief engineer (or equivalent) from the parent contractor's headquarters, the site licensee had veto power over any such appointment.

I–12.10. Control of nuclear related work

The arrangements for the control of nuclear related work included setting up a Jason decommissioning procedure authorization group (JDPAG). The JDPAG approved all nuclear and radiological safety work procedures or method statements covering installation, commissioning, testing, fuel removal, reactor dismantling, repair and maintenance on the Jason reactor and its associated decommissioning facilities. JDPAG membership consisted of DAPs and SQEPs from both the client organization and the contracted resident engineers, together with the radiation protection adviser or operational health physicist from either the fuel removal or reactor dismantling contractor, or both, as appropriate. The JDPAG was chaired by a site licensee staff member and attended by the licensee's reactor engineer, health physicist and health and safety officer.

Most nuclear procedures/method statements were written by the decommissioning contractors and followed a standard format similar to that used elsewhere in the naval nuclear propulsion programme. This format included a title and objective, a list of other relevant reference documents and a list of all DAPs, SQEPs and other operatives involved in the procedure. All plant and system line-ups and other prerequisites were then detailed together with any special precautions that needed to be taken. The nuclear procedure/method statement then detailed all special equipment to be used and contained a job specific hazard and risk assessment. The actual detailed execution of the work was some way into the procedure and all job specific and generic emergency actions that may have been required were at the back of the procedure.

The decommissioning contractors were not necessarily used to this level of detail being prescribed in a method statement and were sometimes not fully conversant with the principle of following its requirements verbatim. Once approved these nuclear procedures or method statements could not be altered in any way without formal change notice approval from the JDPAG. This method of control was applied to all fuel removal and waste removal equipment, activities and operations having direct or indirect nuclear and radiological safety implications. A deliberate policy of separating work prescription from work implementation was adopted throughout the Jason decommissioning project, similar to that used at other naval nuclear propulsion programme related sites, but not at other civilian nuclear sites in the UK.

Considerable time and resources were invested by the contractors to become proficient in the production and execution of the type of method statement required by the site licensee. In general terms, work prescription was provided by the nuclear procedure or method statement, which had to be approved by the JDPAG. Work implementation was effected by adherence to the approved procedure by either the contractors or their subcontractors, as appropriate. Each nuclear procedure or method statement had a nominated procedure co-ordinator who was responsible for the correct sequencing of the work with the individual persons carrying out that work, such as the contractors being responsible for the required standards of workmanship and quality as demanded by the procedure.

I–12.11. Continuity

At the same time as decommissioning, the department was in the process of relocation as outlined above, which imposed additional management constraints over and above those needed to safely manage the decommissioning process. The actual relocation of the department occurred in late October 1998 immediately following fuel removal from the site in September 1998. Prior to fuel removal, considered to be the highest risk activity during the decommissioning project, the site licensee and the full departmental management team were on-site. In recognition of relocation and to aid continuity, the existing operational Jason reactor manager was appointed early on in the project as was the JDS, who would stay permanently on-site until completion of the project.

Notwithstanding the relocation of the department and the senior operational management team, there was clearly a need to maintain an adequate level of site licensee SQEPs to provide continuity and act as a client oversight team throughout the decommissioning project. Three additional dedicated full time chartered engineers, who had previous experience of nuclear, decommissioning and project engineering matters, were recruited and appointed to the Jason operational reactor engineering division prior to POCO. Following relocation, the JDS and his resident decommissioning team were permanently located within the decommissioning facility alongside the decommissioning contractors, acting as the client oversight team. In addition, the JDPAG chairperson was also appointed as the JSAWG chairperson, which conveniently combined the responsibilities for approval and review of all nuclear procedures with those of documents and safety reports forwarded to the full safety committee.

The JDS remained responsible for overseeing and control of work implementation by the various contractors, thereby maintaining the separation between work prescription and implementation. The JDPAG/JSAWG chairperson, the JDS and his resident team were fundamental in providing continuity and maintaining the site licensee's nuclear and radiological safety responsibilities during the middle to later stages of decommissioning when the site licensee and his full management team had relocated.

I–12.12. Records

Jason's operational, maintenance, modification and repair records since installation were very comprehensive and had been scrupulously kept up to date by dedicated operational staff. However, the original reactor installation records (dating from about 1960) and the records of previous nuclear/radiation related operations prior to the installation of Jason were not so comprehensive. The lack of original installation records had the potential to delay the completion of the decommissioning project, particularly during the final stages, when unexpected tritium contamination was discovered that caused difficulty in meeting the site's radiological clearance criterion.

As decommissioning progressed, the contractors made good use of the JDS's knowledge of the reactor layout and its previous operational history. In addition, the whereabouts of retired personnel became increasingly important as they were able to provide the project with very valuable insights and suggestions as the work progressed. This first hand knowledge became crucial during the final stages of decommissioning where unexpected extensive tritium contamination was found in the concrete floors outside the reactor hall, which led directly to a two month delay in completion of the project. This tritium contamination was caused by previous neutron accelerator operations that predated Jason operations. These previous operations came to light primarily through personal contacts with retired personnel rather than through existing operating records. Had this matter been known about or considered earlier, the project delay might have been avoided.

I–12.13. Hold points and authority to proceed

Overall control of sequential packages of work was by means of the hold point and authority to proceed (ATP) system. Hold points were defined early on in the decommissioning project to divide its major work packages or phases into safely manageable packages. The site licensee needed to have a workable safety management hold point and ATP system in place to maintain nuclear and radiological safety standards during the sequencing of decommissioning work packages over and above the traditional work, time and cost based project hold points that occur during any construction/demolition project.

A hold point strategy document was produced that divided up the major work items of the decommissioning project into manageable phases that maintained nuclear and radiological safety, rather than those traditional work based packages that address standard project based time and cost considerations. Each hold point had its own particular completion criteria which would have to be met before the ATP was granted to commence the next work package and proceed to the next hold point. For example, verification of the non-active commissioning of the fuel removal equipment, including completion of defuelling training sessions and the approval of the POSR(FR) was required before the ATP was granted by the site licensee to carry out the actual fuel removal activities.

Within each major hold point there was a series of minor subhold points, with each being cleared by a site licensee staffed acceptance group who would issue an associated subhold point clearance certificate. For example, a subhold point clearance certificate would be issued following correct installation of the fuel removal equipment, and another following testing of the fuel removal equipment prior to carrying out non-active commissioning trials.

The main project hold points were completion of POCO, commissioning of the fuel removal equipment, defuelling, dismantling of the fuel removal equipment, commissioning of the waste handling equipment, reactor dismantling and waste removal, dismantling of the waste handling equipment, radiological clearance of the site and final hand-over of the site. The clearance of the majority of the hold points and subhold points was normally dictated by the progress of work on the ground, but also invariably coincided with approval of the relevant safety case, design report or commissioning schedule.

I–12.14. Permits to work

Day-to-day control of work on the ground during decommissioning had to be significantly upgraded from that in force during previous Jason operations. A new permit to work (PTW) system was adopted, similar to that adopted at other civil UK nuclear sites by the prime contractor, NNC. The large number of personnel carrying out work on-site required the introduction of a robust and comprehensive PTW and certificate of isolation (COI) system, which was required to be in force throughout the decommissioning project to maintain the site licensee's nuclear, radiological and conventional safety standards and responsibilities. It was also required to control the day-to-day work related activities of all personnel working on-site, including the main contractors, subcontractors and client staff.

The existing operational PTW and COI system was significantly upgraded from that used during previous Jason operations. The PTW system covered both nuclear and non-nuclear related activities and was worked up and approved by the site licensee and the prime contractor. It was introduced over a four week trial period alongside the existing permit system until its satisfactory operation was proven.

The overall PTW system consisted of a permit request form signed by a contractor DAP or SQEP, a nuclear procedure (or method statement), a hazard and risk assessment and a COI (for the various electrical and mechanical services that needed to be isolated before the work commenced). The permit had to be endorsed by the client's health physicist and health and safety officer before it was finally approved and formally issued to the contractor by the JDS. The client health and safety officer, who was the single point of contact for all permits, administered the day-to-day workings of the system which included receipt, final issue, monitoring and daily mustering of all permits and certificates in force. All work undertaken within the Jason decommissioning facility required an approved PTW and COI before work could commence, regardless of whether that work was being undertaken by either the decommissioning contractors or client staff.

I–12.15. Transition period activities

The major physical activities that took place during the transition period were POCO and fuel removal. POCO consisted of disabling the reactor and making preparations for fuel removal, including draining all primary, intermediate and secondary water systems and isolating all electrical power supplies to the control rod drive motors, primary coolant pump, reactor control panel and various other reactor control mechanisms and instrumentation systems. The startup neutron source was removed and transferred to Harwell as intermediate level waste (ILW) and many other small redundant calibration and test sources were removed from the site. The reactor hall crane was refurbished and fitted with modern protection devices. A number of unirradiated loose fuel plates were made up into three standard fuel modules which made a total of 16 fuel modules, all of which were stored in the fuel pits. A series of calculations, assessments and measurements were carried out on the reactor fuel, core components, graphite and shield blocks by the contractors, which included taking various drilled core samples in the graphite and shield blocks. The overall aim was to determine potential and measured dose rates and the anticipated extent of the ILW, low level waste (LLW) and free release waste.

Major modifications were made to the building's structure to facilitate fuel removal from the reactor hall and to provide additional emergency exits. This included removing the existing east door of the reactor hall and fitting a new roller shutter door. The local emergency control room arrangements and facilities were also modified and the nuclear safety emergency orders were rewritten to reflect these modifications. The as fitted gamma monitoring systems remained operable throughout the fuel removal phase. The as fitted smoke detection and fire alarm systems were modified and the main access corridors were fitted out with new fire retardant doors. All work on the reactor and associated systems was carried out under nuclear procedural control and in accordance with approved modifications to the operational nuclear safety case. All physical POCO activities were complete by June 1998.

I–12.16. Fuel removal

Fuel had not been transferred from the college for about 25 years and fell outside the scope of the operational nuclear safety case. The fuel removal phase consisted of several interrelated off-site and on-site enabling, preliminary and preparatory activities, including the procurement, preparation and approval of the UKAEA owned Unifetch fuel transfer flask and the design, manufacture, test, installation and commissioning of the equipment needed to install the flask in the reactor hall. The overall aim was to transfer the fuel from the fuel pits to the Unifetch, remove the flask from the reactor hall and subsequently transport it by road to BNFL, Sellafield.

Initial preparatory work for fuel removal off the site consisted of the design, manufacture, assembly and works testing of a purpose built large steel transfer bridge to transfer the Unifetch flask in and out of the reactor hall, and construction of a purpose built flask lid removal gantry and supporting arrangements that were used within the reactor hall. Load and functional testing was done at the manufacturer, witnessed by site licensee staff, the

IPR team and the project management team. The other major off-site work consisted of a full decontamination and maintenance programme on the Unifetch flask, including the removal of all paint from the outside and a full repaint.

Modifications were also made to interface the existing single module Jason fuel flask with the Unifetch flask, as each of the 16 fuel modules had to be moved individually from the fuel pits into the Unifetch. The existing Jason gamma gate, which attached to the bottom of the Jason fuel flask, had to be modified to interface with the Unifetch. A new Unifetch interfacing rotating shield adapter plate was also manufactured, together with new lead shield plugs to seal the individual fuel baskets in the Unifetch once each fuel module was transferred. Trials were also carried out at the manufacturer to functionally test the modified fuel handling equipment using a dummy fuel element. Figure I–35 shows excerpts of the fuel removal operations.

Initial activities enabling fuel removal included carrying out additional radiological surveys and the installation of new decommissioning health physics facilities and equipment. An upgraded ventilation supply to the reactor hall and an extraction system were also fitted, including a mobile filtration unit, new ducting and fire dampers, HEPA filters, an airborne beta monitor, iso-kinetic air samplers and a tritium bubbler.

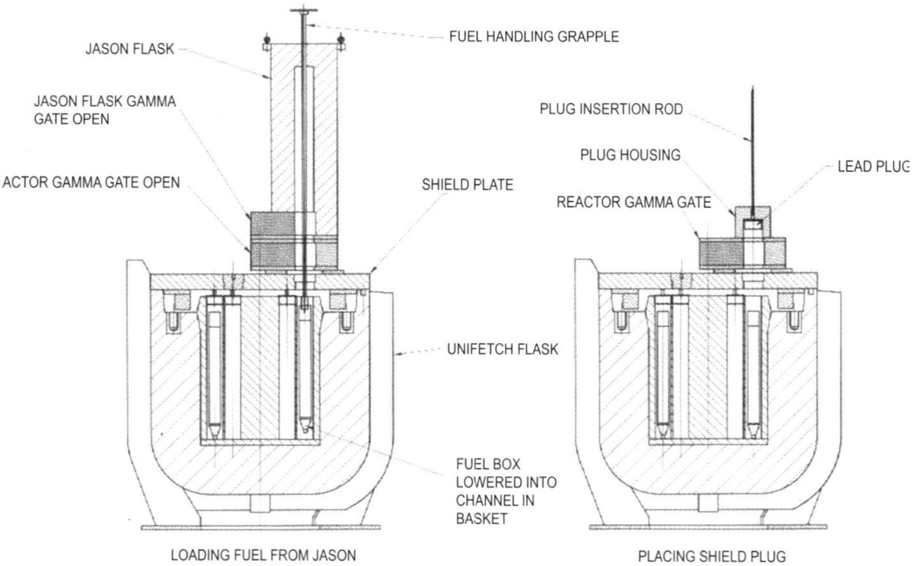

FIG. I–35. Modified fuel removal flask for the Jason decommissioning project.

All preparatory work on-site had to be very carefully managed in order to protect the listed buildings and historical artefacts located underground adjacent to the reactor hall from the effects of the heavy fuel transfer equipment, i.e. cranes, lorries and the Unifetch fuel removal flask. A series of interlocking metal ground protection mats were installed outside the reactor hall to protect the artefacts from the Unifetch flask, the lifting crane and the removal lorry. The basement areas under the reactor hall were also shored up and strengthened by a series of interlocking wooden braces. The steel transfer bridge was erected to transfer the Unifetch flask in and out of the reactor hall via the east roller shutter door (Fig. I–36).

On-site commissioning trials of the fuel removal equipment included load testing of the steel transfer bridge and transferring large test weights in and out of the reactor hall while measuring bridge deflections and clearances, witnessed by both the site licensee's staff and the project management teams. The Unifetch flask lid removal gantry and its support structures were installed within the reactor hall immediately adjacent to the inner end of the transfer bridge. Witnessed operational and load trials, including a statutory proof load test, were also carried out. Non-active commissioning trials were then carried out using the actual Unifetch flask, placing it on the outer end of the bridge, transferring it into the reactor hall, and unbolting and removing the flask lid

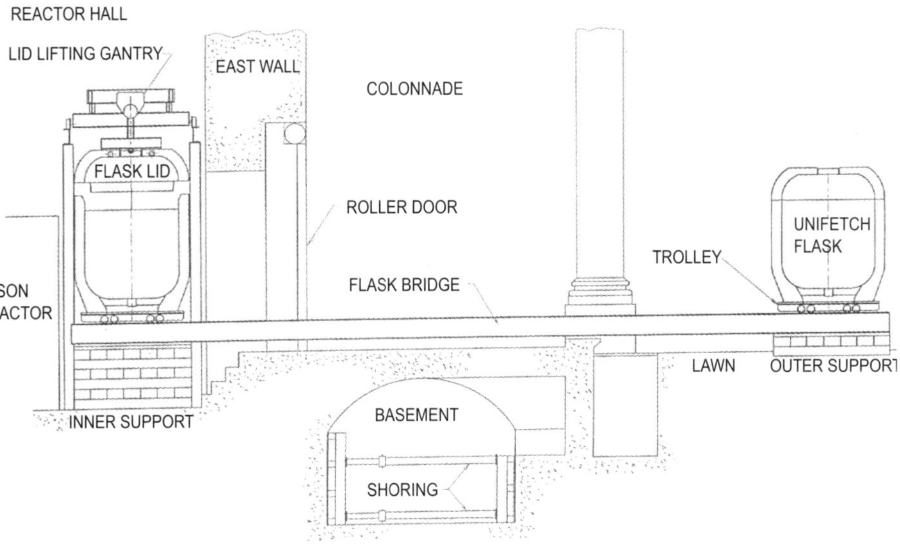

FIG. I–36. Fuel removal from the Jason site.

using the newly constructed gantry. The shield and adapter plate were installed and trials were carried out to gain precise alignment of the small gamma gate, the hole in the shield plate and the selected fuel basket channel in the Unifetch. Further trials were then carried out to confirm the fuel transfer route from the fuel pits to the Unifetch flask and back using a dummy fuel module, the Jason reactor hall crane and the Jason fuel transfer flask.

Additional trials were also made to determine whether a fuel module could be transferred by manual grappling should that become necessary during the defuelling process. The purpose of these trials was to test the equipment and train the main and stand-by contractor fuel removal teams, site licensee and management teams for both normal and emergency situations. The site licensee and the naval regulatory authorities formally required that all the members of the fuel removal teams be SQEPs as a result of previous training and experience and that appropriate training be provided for the specific operation of the Jason fuel transfer equipment. All individuals were required to be competent in specific tasks and to be able to work together as a team. A series of training runs ensured that the correct actions would be taken in the event of an abnormal or emergency situation, either generic (e.g. fire) or specific to refuelling (e.g. dropped load) situation. The overall aim was to ensure that the management and execution of the fuel removal activities would be efficient, effective and safe in all normal and emergency situations. Detailed 'gun drill type' defuelling and emergency procedures were worked up and approved under procedural control to facilitate this requirement.

These internal training sessions were followed by two externally witnessed 'observed runs'. One observed run was carried out to meet the requirements of the MoD(N) headquarters nuclear refuelling safety subcommittee (RSSC) which traditionally witnesses and endorses all practice procedures prior to the actual defuelling and refuelling of a nuclear submarine. These observed runs included emergency situations to test the reactions of the defuelling teams. Formal RSSC acceptance of the observed run was required as part of the overall approval process for the POSR(FR). A second observed run was also carried out and witnessed by members of the Jason reactor safety committee. Following approval of the POSR(FR) on 11 September 1998, the site licensee gave his ATP to remove the fuel, which took two days. The fuel was subsequently transported by road to Sellafield on 16 September 1998.

Similarly to most work related activities, all on-site preparatory and fuel removal activities, including construction, erection, testing, commissioning, training, observed runs, defuelling, flask transfer and fuel removal equipment dismantling activities, were undertaken in accordance with approved permits to work and certificates of isolation, authorized nuclear procedures (or method statements) and approved modifications to the relevant decommissioning safety report (operational safety document, pre-decommissioning safety report (fuel removal) or POSR(FR)).

I–12.17. Radiological protocols

The Jason reactor facility and college site was not licensed under the Nuclear Installations Act but was effectively 'licensed and regulated' by the naval regulatory authority and the chairperson of the naval nuclear regulatory panel (CNNRP). While remaining under MoD(N) control, the site was inspected by the NII under the Ionising Radiations Regulations. The site complied with the governing regulations for the control of exposure to ionizing radiation, transport of radioactive materials, and health and safety in general. Nonetheless, the NII, the Environment Agency (EA) and the CNNRP agreed that the EA would have primacy over the final radiological release of the site for unrestricted future use. Prior to the commencement of the decommissioning project an application was made to the EA for 'approval certificates' to accumulate decommissioning radioactive wastes on-site. The EA also issued similar certificates to dispose of radioactive wastes from the site, covering the relative amounts and respective radioactive nuclide limits for gaseous, aqueous and solid wastes.

Early on in the decommissioning project it was realized that a formal protocol would be required to define the strategy for the final survey of the site and determine the actual radiological site release criterion. In addition to the necessary site cleanup and internal survey, it was also recognized that the final independent radiological survey (IRS) should include the whole site, including the decommissioning facility. It was also considered that this survey should be carried out by an organization of high international and national standing in the radiological field. Numerous meetings took place during the transition period between the NII, EA, CNNRP and the site licensee to work up and agree to the survey and clearance criterion protocols. The agreed protocol stated that the EA would issue 'revocation notices' for the accumulation, storage, discharge and disposal of radioactive substances on the site once they were satisfied that the site had been comprehensively surveyed and met their release criterion, which would signify that the site could be subsequently used with no radiological restrictions. The National Radiological Protection Board (NRPB) was contracted to carry out the IRS according to the agreed protocol. The board would then provide the EA with a comprehensive survey report for each area surveyed. From a technical and public relations point of view, the NRPB was considered to be one of the most suitable, independent, competent and most widely recognized organizations available in the UK to undertake this final radiological survey.

I–12.18. Conclusion

The safety management arrangements, work control procedures and protocols put in place by the MoD(N) to safely manage both POCO and fuel removal during the transition from operation to decommissioning of the Jason reactor helped to maintain the site licensee's nuclear, radiological and conventional safety responsibilities. While some of these arrangements and protocols may be specific to low power reactors, the MoD(N) and the UK, many are generic and can readily be applied to the transition phase in the decommissioning of nuclear reactors or nuclear installations internationally.

I–13. TRANSITION FROM OPERATION TO DECOMMISSIONING OF UKAEA's RESEARCH FACILITIES

I–13.1. Introduction

In the late 1980s and early 1990s the amount of nuclear research and development conducted by UKAEA decreased, particularly in the development of fast (neutron) reactors and the associated fuel cycle technology. Over this period all of its fission reactors were closed as were many of the facilities supporting the research and development programmes.

The role of UKAEA today is:

(a) Restoration of its sites by decommissioning all redundant plants, cleanup of those sites, management of all wastes and, where appropriate, site delicensing;
(b) Fusion research (at its Culham site);
(c) Maintenance of security at its nuclear sites and materials.

Most of the funding for these activities is made available by the Government of the UK (Department of Trade and Industry). UKAEA's mandate includes decommissioning of experimental, prototype and materials test reactors, radiochemical plants, chemical engineering development facilities, materials/waste storage buildings, heavily shielded post-irradiation examination facilities, plutonium handling and processing equipment and a number of particle accelerators. This mix results in a wide variation in the extent of any radioactive contamination and type of hazardous material present (beryllium, sodium, organic solvents, etc.). The facilities date from the 1950s up to the present day. Consequently, particularly for the older facilities, full records are not always available.

The decommissioning strategies are developed for each site and co-ordinated within the overall planning framework covering UKAEA as a whole. In planning the decommissioning, the interactions between facilities and, in a number of cases, between sites are considered, as are the ways the wastes will be managed.

I–13.2. Planning tools

High quality programme, project planning and management were identified early in UKAEA's new role as drivers for the cost effective safe delivery of the restoration programme. Thus, UKAEA developed a suite of programme management tools to help manage this diverse range of liabilities. These include:

I–13.2.1. A parametric cost estimating database (PRICE)

The PRICE is used to assist during the definition, planning and initiation phases of decommissioning projects. The database is updated with feedback from the tendering process and the eventual project costs. The system is used to assist programme and project managers in making the case for adequate financial provisions as an input to decommissioning option studies and option selection.

I–13.2.2. A care and maintenance guidance document [I–34]

This guide (in conjunction with a workbook created with ACCESS) is used to develop the most appropriate care and maintenance strategy for redundant facilities. The system consists of a set of standard principles which, together with the workbook and a questionnaire, guide the planner in a manner that ensures that the appropriate safety and environmental standards are maintained together with cost effective solutions.

I–13.2.3. A programme prioritization methodology

In order to manage within a range of constraints, UKAEA prioritizes at the project and sub-project level within the overall decommissioning and waste management programme. A multi-attribute analysis derives the prioritization ranking, recognizing that human and financial resources do not permit all work to be carried out immediately. In determining the final priority, however, it is also necessary to recognize the constraints that arise from intra-programme and inter-site dependencies.

I–13.2.4. Strategic planning system software (SPS)

The SPS is a decision support software package which models the decommissioning and waste management programme and flows. It is used to assist with the development of the overall site restoration programme and to provide an auditable estimate of the total liability. The principal outputs are costs, timings and utilization data for processes, waste storage and disposal facilities. The SPS assists with decommissioning option studies and the production of an optimized programme strategy.

I–13.2.5. Programme risk assessment and management

The nature of the restoration programme is such that there are inherent risks at both the project and programme level. Such risks arise from a number of sources ranging from the technical risk associated with the history of the work undertaken to the socioeconomic or sociopolitical risk. However, provision has been made in the programme to recognize and, where practical, manage these risks. Assessment and control arrangements are included to ensure that the risks involved in the decommissioning and waste management processes are properly managed.

I–13.3. Overview of AECP 1085

Government policy in the UK is that the process of decommissioning redundant plants and facilities is to be undertaken “as soon as it is reasonably practicable to do so, taking account of all relevant factors” (Cm 2919, para. 124). The policy of UKAEA on the timing of decommissioning requires that the first phase of decommissioning be carried out immediately following shutdown, with the timing of subsequent stages depending on individual circumstances, interdependencies with other facilities, safety priorities and the economic benefit to be gained from accelerated work.

In general, the process of decommissioning redundant UKAEA plants and facilities is begun as soon as it is reasonably practical to do so and on a timescale that ensures the systematic and progressive reduction of hazards. For some facilities there may be planned hold points (in some cases for a number of years) to improve overall safety and reduce the environmental impact of the decommissioning process. During these periods the facility is kept in a ‘care and maintenance’ state until the next stage of decommissioning. AECP 1085 provides guidance on the development of a cost effective care and maintenance plan. The guidance can be applied to any chosen starting point (e.g. post-operational, post-first phase, etc.). Where little or no POCO or safety preparation has been carried out, the document can be applied to different options as a tool to help define the optimal level of POCO and safety preparation. The main steps in applying the guidance document are given in the following sections.

I–13.3.1. Establishment of the facility's 'baseline' status

The baseline status is that of the facility at the point chosen to apply the guidance document (i.e. the start point for the next phase of work). Baseline information will typically include:

(a) A list of all operating systems and equipment (including safety related equipment) in the facility (e.g. ventilation systems, fire detection equipment, water, gas and electricity supplies, operating rules and instructions, etc.);
(b) The current maintenance schedules and arrangements;
(c) Inventories of radiological and hazardous materials;
(d) The status of all structures, outbuildings, effluent delay tanks, etc.;
(e) Staffing requirements;
(f) Current environmental monitoring regimes in workplaces.

I–13.3.2. Carrying out of a preliminary hazard assessment

A preliminary hazard assessment is carried out to identify the areas where physical and managerial safety systems will be required to control the hazards that will be present during care and maintenance. Where POCO has not been undertaken the preliminary assessment will also help to identify those hazards that should be removed during this operation.

I–13.3.3. Assignment of approximate costs and drivers to each baseline system

The cost and driver for each baseline system is established. Typical drivers include safety/legislative, people/occupancy, corrosion prevention, contamination control, environmental protection, etc.

I–13.3.4. Prioritization of systems for detailed analysis

Each of the facility's systems is prioritized against cost and the applicability of the drivers under a care and maintenance regime is questioned. Each (prioritized) system is further analysed to identify whether it can be modified, reduced or removed altogether. All decisions must ensure safety and compliance with legislation.

I–13.3.5. Submission for formal hazard assessment

The formal hazard assessment will form part of the safety case covering the care and maintenance period.

I–13.3.6. Preparation of the final report

A suggested format of the final report is meant to ensure consistency of the care and maintenance plans for different facilities and sites. This report is prepared by a suitably qualified and experienced team. Typically a team comprises the facility manager, the maintenance manager, a radiological protection advisor, and experts in safety, civil engineering, ventilation, corrosion and project risk assessment. The following sections give two examples of the transition from operations to decommissioning.

I–13.4. Materials testing reactors

The DIDO and PLUTO materials testing reactors operated at Harwell until their closure in March 1990 (Fig. I–37). They both operated for over 30 years at various powers up to a maximum of 26 MW. The reactor cores were designed to produce very high neutron fluxes which enabled the effects of long term irradiation on materials to be tested in accelerated timescales.

The first phase of decommissioning commenced immediately after shutdown. This involved routine operations to defuel the reactors and drain the heavy water. The second phase started after approval of the decommissioning safety case. Initial operations concentrated on the removal and disposal of the test rigs having the highest inventory and the consolidation of the remainder into shielded storage blocks. In addition, contaminated systems and unwanted equipment were progressively cleared from the reactor containment buildings.

Studies showed that the existing ventilation systems were ageing and were relatively expensive to maintain and operate. Therefore new systems were installed in 1995. These now control both temperature and humidity to prevent condensation and minimize corrosion within the core containment buildings. New electricity supplies meeting modern standards were installed. These supply only essential services (low voltage lighting, ventilation plant, monitoring equipment, etc.).

Care and maintenance is now based upon assessments made in accordance with AECP 1085. Both the radiological and environmental monitoring regimes have been modified to reflect the hazard reductions achieved during the transition from operation to care and maintenance.

The reactor core containment buildings no longer require permanent occupation but are fitted with monitoring systems that alert staff elsewhere on the site if problems should arise. Many of the peripheral buildings have, where possible, been decommissioned and demolished to 'brown field' status.

I–13.5. Decommissioning of building 351

Building 351 was a seven storey chemical engineering research facility located at Harwell. In 40 years of operation it was occupied by many hundreds of test rigs and pilot scale experimental plants, and housed dedicated ventilation and liquid effluent delay systems. At its peak, over 200 R&D staff populated the building, working on all aspects of the nuclear fuel cycle from fuel fabrication studies to waste vitrification process development. As requirements changed there was less need for facilities of this type and building 351 was declared redundant in 1990. Unwanted equipment from earlier years remained in over one hundred radiologically designated areas throughout the building.

Initial analysis of possible strategies for the building included a thorough examination of a variety of scenarios. This established that a steady decommissioning programme followed by demolition of the building was considerably cheaper than delayed decommissioning after a period of care and maintenance. Early decommissioning planning involved establishment of baseline information. This included surveying each area again and removing samples for analysis to reduce the uncertainties regarding the radioactive inventory and the nature of the chemicals present. Waste volume estimates and disposal routes were also established using the information generated. Operating and maintenance records were used as a source of data for preliminary safety case assessments. Knowledge was also built up through interviews with plant operators and health physics staff, and by searching archive records.

The POCO operations proceeded in parallel with planning to further reduce uncertainties and hazards. The planning phase was also used to identify and test the equipment required for the decommissioning phase. Efforts concentrated on identifying appropriate levels of technology to solve specific problems that had not arisen during operation or maintenance. Wherever possible equipment (such as cutting tools, temporary containment and ventilation systems, remote viewing and handling equipment) that could be used on more than one rig or plant was specified. Redundant services and supplies were isolated and removed where possible. Electricity to essential equipment and lighting was supplied by new cables installed to aid the identification of 'live' power supplies during decommissioning operations. The main

FIGURE I–37 LEGEND

(1) Reactor aluminium tank containing heavy water
(2) Heavy water level
(3) Fuel element
(4) Vertical experimental hole thimble
(5) Control signal arms (six)
(6) Experimental hole
(7) Top biological shield
(8) Graphite reflector
(9) Experimental holes
(10) Lead thermal shield (water cooled)
(11) Reactor steel tank with boral sheet lining
(12) Experimental hole entering heavy water zone
(13) Experimental hole entering graphite zone
(14) Vertical experimental hole entering graphite
(15) Concrete biological shield
(16) Reactor supporting structure
(17) First floor
(18) Thermal column
(19) Hollow stanchions communicating with base of reactor
(20) Steel top plate and ring
(21) Duct for experimental services
(22) Biological shield casing plates
(23) Ion chamber
(24) Fine control rod gear box
(25) Heavy water outlet pipe
(26) Heavy water storage tank
(27) Heavy water dump tank
(28) Heavy water heat exchangers
(29) Heavy water main circulation pump
(30) Secondary cooling water piping (to cooling towers)
(31) Heavy water emergency circulator
(32) Experimental cooling system

FIG. I–37. Technical features of the DIDO reactor.

decommissioning programme commenced in 1993/1994 and culminated in the demolition of the building's superstructure in 1997.

I–13.6. Conclusions

Successful decommissioning depends on thorough planning and a site restoration programme. Such planning has to take constraints and motivators into account, particularly the interdependencies between projects and sites. Despite thorough planning, some inherent risks will remain. Their identification, assessment and management is an essential part of the programme management function.

Establisment of facility baselines is essential and should involve collation of information from all available sources. This should also be done by a team of people with relevant technical expertise and appropriate experience. Assessments made during the transition period can identify the appropriate levels of technology to solve decommissioning problems and the equipment that can be used for more than one application. Use of a number of specialized planning tools, the involvement of a knowledgeable team of personnel and the use of appropriate programme and project management techniques will ensure the success and cost effectiveness of the restoration programmes.

I–14. THE TRANSITION TO DECOMMISSIONING IN THE USA

I–14.1. Introduction

Post-shutdown, pre-decommissioning experience in the USA can be differentiated between two population groups of facilities, US Department of Energy (USDOE) facilities and NPPs. In addition to these two population groups there are also university owned research reactors, but as these are well in the minority they are not discussed further here.

The USDOE is faced with many excess facilities that will eventually be decommissioned. The department has been establishing and applying systematic methods for stabilizing and placing such facilities in a deactivated condition with caretaker surveillance and maintenance. Details of this experience are readily accessible on the Internet.

In the case of NPPs, what once was perceived by many to be an increasing trend toward decommissioning has abated. Therefore, there is limited interest in NPP decommissioning in the USA. No closures are currently planned aside from those already in process. Aside from these, the Electric Power Research Institute has the most active NPP decommissioning technology programme.

I–14.2. Distinguishing differences between decommissioning programmes

A variety of differences between USDOE facilities and commercial NPPs result in a somewhat different situation with respect to decommissioning programmes and planning.

I–14.2.1. Population characteristics of USDOE facilities

There are several thousand government owned and USDOE operated facilities, of which 100–200 are major facilities and the rest are of lesser significance. Among these, there is a wide variability in:

(a) The types of facility and their purpose — reactors, chemical process, radiation experimentation, industrial processes, storage of materials, fuel, and waste — with missions such as nuclear energy research and development for fission, fusion, propulsion, isotopic generators, isotope production, materials development, weapons programmes and others;
(b) The age of the facilities and design and construction codes dating back to the 1940s;
(c) Substantial physical degradation in many cases, depending on when the facilities were last used;
(d) Types and degree of contamination (many facilities have some degree of alpha contamination);
(e) Types of hazardous materials such as metallic sodium, acids, solvents, etc.

I–14.2.2. Population characteristics of US NPPs

Approximately 100 PWR and BWR NPPs are owned and operated by electricity suppliers. These are characterized by:

(a) Singularity of mission — the generation of electricity;
(b) Similarity in design vintage (generally 25 years or younger);
(c) Very robust construction with very little degradation in the nuclear steam supply system and associated systems;
(d) Predominantly beta–gamma contamination of almost all activated corrosion products;
(e) Almost no hazardous chemicals — primarily for water treatment — and thus very little mixed waste.

I–14.2.3. Differences in programmatic characteristics

Significant management and programmatic differences between USDOE facilities and NPPs affect the approach to and current status of decommissioning, specifically:

(1) Lifetime momentum: Many government programmes of nuclear related research are on the wane; indeed much research is focused on decommissioning technologies and demonstrations. In contrast, NPPs have a steady, identifiable, constant mission for which continuation has been enhanced by:
   (i) The recognition that reserve margins for electricity in the USA are quite low,
   (ii) Licence renewal rules that allow 60 years of operation.

(2) Regulation: Regulation of government facilities is internalized for the most part within independent government oversight organizations. Government regulation of the transition to decommissioning is, by necessity, functional in nature because of the wide variation in facilities. Nuclear power plants are regulated by an independent agency, the US Nuclear Regulatory Commission, which may be prescriptive as needed because of the relative sameness of the NPPs.

(3) Funding: Decommissioning government facilities is self funded and generally budgeted on a year to year basis, competing with research and development funds. Decommissioning trust funds are established for NPPs and funded by electricity users over the lifetime of their operation.

Therefore, while some technological aspects such as ALARA, decontamination, remotely operated demolition and size reduction equipment, analysis methods, etc. may be applicable to both USDOE facilities and NPPs, the details of management and planning for decommissioning vary considerably between them.

I–14.3. The USDOE's position

The USDOE has extensive experience in the transition of facilities from an operating mission to permanent shutdown. The transition process is becoming more systematic for many of the reasons discussed in this report. This process is embodied in USDOE Orders and Guides that form the regulatory bases that govern the end of a facility's life cycle. These are:

- DOE Order 430.1A: Life cycle asset management;
- DOE G 430.1-2: Implementation guide for surveillance and maintenance during facility transition and disposition;
- DOE G 430.1-3: Deactivation implementation guide;
- DOE G 430.1-4: Decommissioning implementation guide;
- DOE G 430.1-5: Transition implementation guide.

Once a facility is no longer needed a decision is reached as to whether or not it is to be promptly decommissioned, which rarely happens. The alternative is to place the facility in a deactivated condition, which is defined as the process of placing a facility in a stable and known condition, including the removal of hazardous and radioactive materials to ensure adequate protection of the workers, public health and safety, and the environment, thereby limiting the long term cost of surveillance and maintenance. Actions include removing the fuel, draining and/or de-energizing non-essential systems, removing stored radioactive and hazardous materials, and related actions. Deactivation does not include all the decontamination necessary for the dismantlement and demolition phase of decommissioning, e.g. removal of contamination remaining in the fixed structures and equipment after deactivation.

Deactivation is similar in concept to SE. The major difference is that many USDOE facilities do not have containments as reactor plants do, so 'enclosure' does not apply in many cases. The key to deactivation is to ensure that remaining hazards and contamination in a facility are controlled and immobile.

I–14.4. USDOE transition planning: Example 1, transfer of facilities to D&D

In the past, some facilities were declared redundant and essentially abandoned without having been completely prepared for an extended period of non-operation. While such facilities do not pose any uncontrolled safety or environmental risk, this practice can result in higher decommissioning costs. As a result, transfer of such facilities to D&D would result in expenditure for actions that should have been taken earlier at lesser cost.

An approach was developed that is philosophically the same as purchasing a house and having it inspected. A process was developed and tools such as checklists and standard report formats were developed. Figure I–38 illustrates the process used to transfer a facility from an operating organization to one that is to maintain it in an inactive condition. Table I–VII is an example of a top level checklist of the actions that are used for such a transfer.

Complex facilities have been surveyed, pre-transfer conditions specified, and post-transfer surveillance and maintenance defined. The results, when converted into sufficiently detailed completion specifications (called end

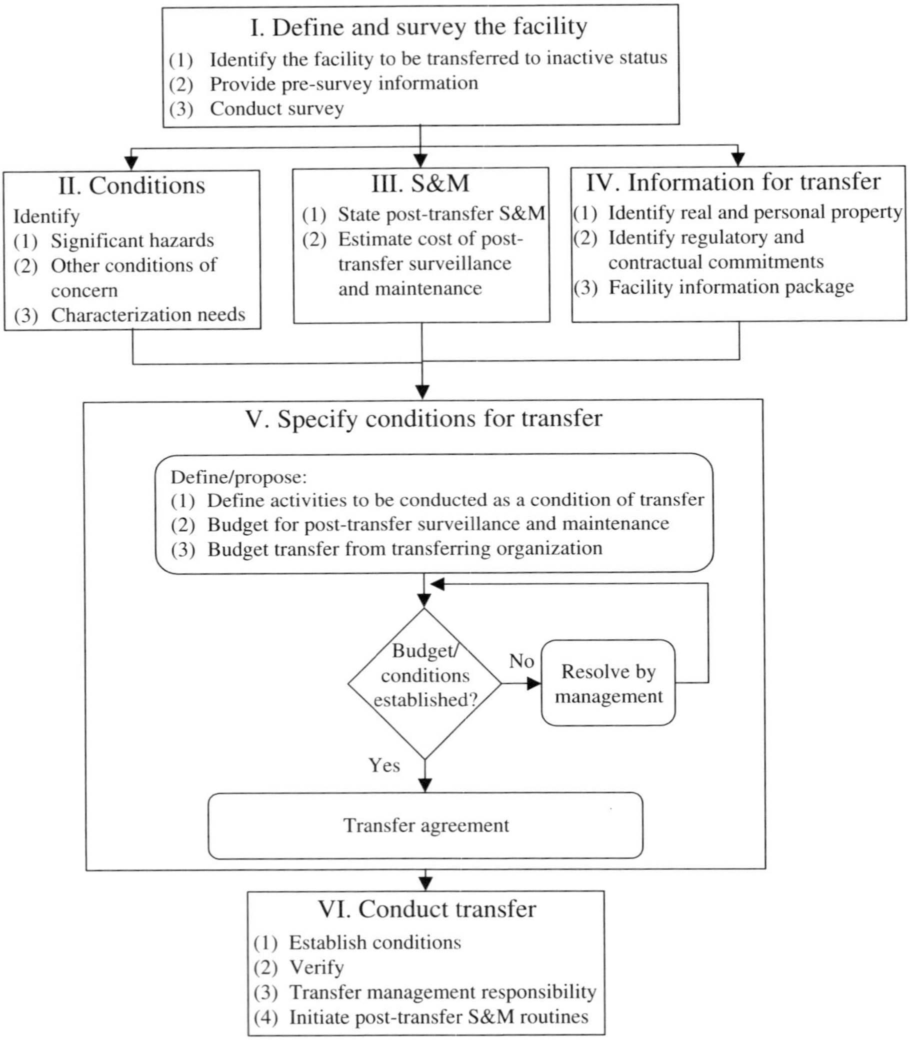

FIG. I–38. Steps in the transition of a redundant facility to D&D.

TABLE I–VII. TRANSFER CHECKLIST

Activity	Functions
Decision to proceed, planning	The facility is declared redundant and a request is initiated for transfer to inactive status. A schedule of activities is identified using this checklist. Interface contacts are identified in the transferring and recipient organizations.
Determine physical boundaries/transfer scope	Identify all structures, outbuildings, tanks, etc. that should be included in the transfer. Identify any issues related to separation of systems that serve other facilities that are not to be transferred.
Conduct a facility survey	Determine the condition/status of structures and systems. Identify what characterization is needed. Identify the necessary pre-transfer operations.
Cost evaluation	Estimate the post-transfer surveillance and maintenance (S&M) budget using a task basis.
Budget	Take the necessary budget actions.
Staffing	Identify staff that know the facility and would contribute to continuity for deactivation or decommissioning. Where possible, arrange for reassignment.
Identification of actions to be tracked prior to transfer as well as other commitments that will be assumed by the recipient	Generate a list. For example, facility repairs to be completed, pre-transfer end points, contracts and purchase orders that need to remain in place, etc.
Permits, licences, agreements, safety analysis, stakeholder commitments, etc.	Identify commitments that will remain after transfer and understand the actions needed (e.g. regulatory agreements with State authorities which have jurisdiction over hazardous emissions).
Inventory of nuclear and fissionable materials	Obtain a listing. Unless otherwise agreed, removal would be the responsibility of the transferring organization which would record the results.
Toxic, hazardous and radioactive materials	Obtain a listing. Determine what is to be removed by the transferring organization (based on justifiable criteria or as otherwise mutually agreed) and record the results.
Characterization	Summarize the radiological and hazardous chemical contamination conditions that exist.

TABLE I–VII. TRANSFER CHECKLIST (cont.)

Activity	Functions
Authorization basis, safety analysis, operating specifications	Report the status, list the defining documents, and identify the S&M requirements for maintaining the safety envelope. Update or revise as appropriate for completion of stabilization activities.
Safe shutdown implementation plan (if applicable)	Verify safe shutdown needs. Address the configuration and contamination conditions that should be established prior to transfer.
Property assets	List excess assets that are of significant value. Identify property that should be transferred.
Schedule integration	At the conceptual level, develop a schedule that integrates completion of operations and stabilization with post-transfer S&M, initiation of deactivation and/or decommissioning.
Pre-transfer operations	For stabilization and hazard elimination activities, agree and conduct as a condition of transfer.
Formal documents for conducting the transfer	Draft, negotiate, revise and finalize transfer memoranda and agreements as decided.

points, see next section), provide an excellent basis for establishing a project plan and subsequent work plans for stabilizing a facility.

I–14.5. USDOE transition planning: Example 2, creating objective driven project management for deactivation of facilities

Often, managers and engineers who operate facilities do so in an ongoing production mode. When they are assigned the job of shutting down and deactivating a facility their mind set must be changed to a project management approach. In addition, there is a tendency to do too much, for example complete decontamination when this cannot be achieved within budget or schedule.

The term ‘end point’, which was first used for facility retirement during the cleanup of Three Mile Island Unit 2, is commonly applied and systematic methods have been derived to determine what they should be. End points for deactivation are similar in concept (but much different in reality) to construction specifications. That is, end points are specifications of the conditions to be achieved in each room and for each system and major piece of equipment in a facility. Two methods were developed for deriving end points. One uses computer software and end points that are hierarchically derived. Another consists of a set of checklists, the subjects of which are based on experience. It should be noted that a large number of end points can be derived for a complex facility, but many can be achieved within a single work package, for example in a common physical area.

Creating end point specifications for deactivating facilities has worked extremely well. They provide detailed requirements readily understood by engineers and workers responsible for establishing the conditions, thus supporting development of work packages, schedules, and budgets. A second major value is that the level of detail demonstrates to interested parties who are not involved in the day-to-day work that the deactivation job will be thorough and result in a stable, well managed condition. Results achieved with large facilities have been cutting years off originally conceived deactivation schedules, which in turn results in staff reductions in the hundreds and the associated savings.

I–14.6. USDOE transition planning issue: Example 3, establishing use of methods across the DOE complex

Over the past 5 to 10 years, deactivation and stabilization activities have created several methods and tools, including those described above. In order to make them available to planners and engineers at the many facilities across the DOE sites, several hundred pages and methods, as well as directly related field experience have been made available on DOE-EM's web site. These can be accessed with the following addresses:

Subject	Address
Excess facility transfer to deactivation and decommissioning — overview and navigation map to the web site and related ones such as the links below	http://www.em.doe.gov/deact
Facility survey and transfer	http://www.em.doe.gov/deact/fst.html
Project management plans	http://www.em.doe.gov/deact/pmp.html
Deactivation management	http://www.em.doe.gov/deact/dm.html
Deactivation completion	http://www.em.doe.gov/deact/comp.html

I–14.7. US NPP transition planning issue: Conducting sufficient pre-shutdown activities

The Electric Power Research Institute, which provides service to many US NPP owners, recognizes that a logical phased approach to decommissioning planning and scheduling is needed. Activities are listed in five decommissioning phases, of which Phases I and II below are relevant to transition.

I–14.7.1. Phase I: Planning and preparing for decommissioning

This period is considered as the initial planning period up to plant shutdown. Planning activities in this phase included the following:

(a) Planning assumptions and design bases,
(b) Strategic and project plans,
(c) Technical and feasibility studies to support projects,
(d) Schedules,
(e) Modifications to processes,
(f) Key decisions and milestones,
(g) Personnel transition plans,
(h) Licensing submittals,
(i) Plant design modifications,
(j) Cost estimates,
(k) Contingency plans,
(l) Low level waste plans,
(m) Spent fuel management,
(n) Site characterization,
(o) Communication with employees and stakeholders,
(p) Labour agreements.

I–14.7.2. Phase II: Preparing and modifying the plant for D&D

Phase II begins at the time of plant shutdown and continues until major D&D commences. During this phase, design modifications are made to the plant to prepare it for major D&D. Tasks during this phase include:

(a) Asbestos removal;
(b) Characterization and disposition of operating systems;
(c) Chemical decontamination of the primary systems;

(d) Installation of decommissioning plant modifications such as alternate power systems, alternate fuel pool cooling systems, monitoring station, alternate radiation monitoring systems;
(e) Installation of temporary facilities;
(f) Installation of radwaste processing systems;
(g) System tagouts and draining.

I–14.7.3. Phases III, IV, V and beyond

The remaining phases are beyond the scope of this report but bear on phases I and II:

- Phase III: Decontamination and dismantlement. This phase includes removal and shipment of hardware for off-site disposal, decontamination of remaining components, and dismantlement of plant structures.
- Phase IV: Site release and licence termination by the NRC. Following dismantlement of plant structures, additional site characterization is necessary to ensure that the site may be released to unrestricted release in accordance with regulations. The licence termination plan must be submitted to the NRC two years prior to the termination date of the plant licence. Once site release criteria have been met and confirmatory surveys completed, the plant's NRC licence may be terminated.
- Phase IV and beyond: Dry spent fuel storage. This is the period following release of the protected area of the site through removal of all spent fuel from the site.

I–14.8. Conclusions

In the USA, with a few exceptions, the current emphasis of the commercial utility industry is not on decommissioning, but on power generation and, in many cases, obtaining licences to extend the operational lives of their NPPs to 60 years. The primary purpose of planning the future transition is to provide decommissioning cost estimates as a basis for defining financing requirements.

The USDOE on the other hand is faced with the prospect of decommissioning several hundred major nuclear and radiological facilities and thousands of minor ones that no longer have a mission. Many of these have been shut down for years. The cost of their combined decommissioning is spread over 20 years or more by budgetary constraints. These facilities fall into three categories:

(1) Priority projects for major facilities to prepare for and conduct decommissioning;
(2) Facilities that are actively maintained in a shutdown state but which cause enough hazards to require continuous manning;
(3) Many facilities that are in a caretaker or abandoned mode with minimal or no operating equipment and with the primary emphasis on controlling contamination. The USDOE has established guidance for transition that is consistent with the principles and methods in this report.

REFERENCES TO ANNEX I

[I–1] UHLMANN, P., "Decommissioning of nuclear facilities: Discussion of social consequences", Safe Decommissioning for Nuclear Facilities (Proc. Int. Conf. Berlin, 2002), IAEA, Vienna (2003) 411.

[I–2] KOGLIN, D., LOESCHORN, U., "Technical aspects of the decommissioning and the dismantling of nuclear facilities with special consideration of the Greifswald NPP", Practical Decommissioning Experience with Nuclear Installations in the European Community (Proc. 3rd Seminar Luxembourg, 1993), Commission of the European Communities, Luxembourg (1993).

[I–3] OECD NUCLEAR ENERGY AGENCY, The NEA Co-Operative Programme on Decommissioning — The First 10 Years 1985–95, OECD, Paris (1996).

[I–4] WON-ZIN OH, GYE-NAM KIM, HUI-JUN WON, "Studies on decommissioning of TRIGA reactors and site restoration technologies in the Republic of Korea", Decommissioning Techniques for Research Reactors, IAEA-TECDOC-1273, Vienna (2002).

[I–5] INTERNATIONAL ATOMIC ENERGY AGENCY, Principles for the Exemption of Radiation Sources and Practices from Regulatory Control, Safety Series No. 89, IAEA, Vienna (1988).

[I–6] OECD NUCLEAR ENERGY AGENCY, INTERNATIONAL ATOMIC ENERGY AGENCY, EUROPEAN COMMISSION, Nuclear Decommissioning — A Proposed Standardised List of Items for Costing Purposes, OECD, Paris (1999).

[I–7] BARSEBÄCK KRAFT AB, OKG, SYDKRAFT, Kärnkraften svarar för en betydande del av Sydkrafts elproduktion (Nuclear Power Contributes a Significant Proportion of Sydkraft's Electricity Production), Barsebäck Kraft, Okg Sydkraft (1999).

[I–8] SWEDISH NUCLEAR POWER INSPECTORATE, Kärnkraftens Historia (The history of nuclear power), www.ski.se, SKI, Stockholm (2001).

[I–9] KLUGE, J., Framtidsfabriken (The factory of the future), Barsebäck Kraft AB, Löddeköpinge (1999).

[I–10] WESTESSON, K., Nya Fabriken (The new factory), Barsebäck Kraft AB, Löddeköpinge (2000).

[I–11] SWEDISH GOVERNMENT, En förnyad Prövning av Stängningen av Barsebäck 2 (Re-examination of Shutting Down Barsebäck 2), Rep. Skr 2001/02:22, Ministry of Industry, Employment and Communications, Stockholm (2001).

[I–12] SWEDISH GOVERNMENT,, SYDKRAFT; VATTENFALL, Avtalssamling — Uppgörelsen mellan Staten, Sydkraft, Vattenfall om en förtida stängning av Barsebäck 1 (The terms of the Agreement Between the Swedish State, Sydkraft, Vattenfall on Prematurely Shutting Down Barsebäck 1), Sydkraft Service Partner AB, Malmö (1999).

[I–13] SYDSVENSKA VÄRMEKRAFT AB, BARSEBÄCKKRAFT AB, Tillämpningsbestämmelser Angående Kvalitet och Säkerhet Samt Avställning- och Servicedrift vid Barsebäcksverket (Quality Security and Service Operations at Barsebäck), SVKAB/BKAB, Malmö (2000).

[I–14] SWEDISH GOVERNMENT, Kärntekniklagen (The Nuclear Matters Law), 1984:3, Ministry of the Environment, Stockholm (1984).

[I–15] SWEDISH GOVERNMENT, Strålskyddslagen (The Radiation Protection Law), 1988:220, Ministry of the Environment, Stockholm (1988).

[I–16] SWEDISH GOVERNMENT, Finansieringslagen (The Nuclear Fuel Liabilities and Financing Law), 1992:1537, Ministry of the Environment, Stockholm (1992).

[I–17] SWEDISH NUCLEAR FUEL AND WASTE MANAGEMENT COMPANY, Om Sveriges Radioaktiva Avfall (Radioactive Waste Management), SKB, Stockholm (1997).

[I–18] SWEDISH NUCLEAR FUEL AND WASTE MANAGEMENT COMPANY, FUD-Program 2001 (Radioactive Waste R&D, Handling and Storage), SKB, Stockholm (2001).

[I–19] SWEDISH GOVERNMENT, Miljöbalken (The Environmental Code), Ministry of the Environment, Stockholm (1999).

[I–20] SWEDISH NUCLEAR POWER INSPECTORATE, Om SKI (The Inspectorate), www.ski.se, SKI, Stockholm.

[I–21] SWEDISH RADIATION PROTECTION AUTHORITY, Myndigheten SSI (The Authority), www.ssi.se, SSI, Stockholm.

[I–22] SWEDISH ENVIRONMENTAL PROTECTION AGENCY, Om Naturvårdsverket (The Agency), www.environ.se, Swedish EPA, Stockholm.

[I–23] SWEDEN'S COUNTY ADMINISTRATIONS, Länsstyrelsens Roll (The Administration), www.lst.se, Sweden's County Administrations, Stockholm.

[I–24] SWEDISH NUCLEAR POWER INSPECTORATE, Statens Kärnkraftsinspektions Föreskrift om Säkerhet i Vissa Kärntekniska Anläggningar (Nuclear Regulation Matters in Certain Nuclear Facilities), Rep. SKIFS 1998:1, SKI, Stockholm (1998).

[I–25] BISSMARCK, G., Strategisk Plan för avvecklingen av Barsebäck 1 (Strategic Decommissioning Plan), Rep. BF-200107-11, Barsebäck Kraft AB, Löddeköpinge (2001).

[I–26] PÅLSSON JAN, O., Verksamhetsrapport BFA — 2000 (Activity report), Rep. F-200103-03, Barsebäck Kraft AB, Löddeköpinge (2001).

[I–27] WESTESSON, K., Verksamhetsrapport FP — 2000 (Activity report), Rep. F-200103-02, Barsebäck Kraft AB, Löddeköpinge (2001).

[I–28] PÅLSSON JAN, O., Verksamhetsrapport BFA — 2001 (Activity report), Rep. BF-200202-05, Barsebäck Kraft AB, Löddeköpinge (2002).

[I–29] WESTESSON, K., Verksamhetsrapport 2001 för enhet BFP (Activity report), Rep. BF-200202-18, Barsebäck Kraft AB, Löddeköpinge (2002).

[I–30] HANSFORD, E.W., FRANCIS, R.J., BROADBENT, J.H., "Jason Decommissioning Project", paper presented at the IBC Decommissioning Conf. London, 1998, IBC UK Conferences Ltd, London (1998).

[I–31] HOULT, D., HANSFORD, E.W., "Decommissioning the Jason reactor inside an ancient monument", paper presented at the IBC Decommissioning Conf. London, 1999, IBC UK Conferences Ltd, London (1999).

[I–32] INTERNATIONAL ATOMIC ENERGY AGENCY, Decommissioning of small Medical, Industrial and Research Facilities, Technical Reports Series No. 414, IAEA, Vienna (2003) 175.

[I–33] LOCKWOOD, R.J.S., BEELEY, P.A., BARRATT, K.L., "Decommissioning the Jason low power research and training reactor at Greenwich — An overview of the project, the nuclear safety management arrangements and important lessons learnt", paper presented at the IBC 7th Int. Conf. on the Decommissioning of Nuclear Facilities, London, 2000, IBC UK Conferences Ltd, London (2000).

[I–34] UNITED KINGDOM ATOMIC ENERGY AUTHORITY, Care and Maintenance Guidance Document, Rep. AECP-1085, UKAEA (1998).

Annex II

PROBLEMS ENCOUNTERED AND LESSONS LEARNED FROM THE OPERATION TO DECOMMISSIONING TRANSITION PERIOD OF SELECTED NUCLEAR FACILITIES

The following examples of lessons learned comprise brief technical information on the nuclear facilities involved and an outline of the problems encountered. The situations described are typical of the difficulties that can arise when planning or implementing activities typical of the transition period. Although the information presented is not intended to be exhaustive, the reader is encouraged to evaluate the applicability of the lessons learned to a specific decommissioning project.

II–1. TRINO NPP (ITALY) — CASE No. 1

II–1.1. Problem

The decontamination of non-irradiated fuel elements which had been placed in the core between 1987 and 1992 (the date when the core was discharged) but never exposed to neutron flux provided the opportunity to carry out a series of checks on the elements when they were removed from the spent fuel pool, decontaminated, fuel dried and transferred to the fresh fuel storage facility after having been placed in a nylon container (Fig. II–1).

During the removal, surveys were conducted to ascertain the presence of hot spots due to crud deposits on the element. This showed, in certain cases, a dose rate that was more than twice that anticipated.

II–1.2. Solution found

The high dose rate was due to metal shavings on the elements, originating from 1968 when the thermal shield was cut. As it was impossible to remove those shavings using the available equipment, a purpose built nozzle with appropriate adaptors was produced at the plant. This made it possible to direct the water flow towards the shavings and remove them by flushing. This operation was completed very quickly except for one case which took several hours. In this way the contractual conditions relating to fuel dose rates for the sale of the fuel elements could be met.

II–1.3. Lesson learned

The lesson learned was that even the simplest operations can lead to unforeseen problems originating from previous and almost forgotten operations. Hence it is necessary to be prepared for all eventualities.

II–2. TRINO NPP (ITALY) — CASE No. 2

II–2.1. Problem

Prior to the dismantling of plant components, a number of preparatory activities are required to remove dangerous materials and reduce health risks to the workforce. At Trino, this included removal of the thermal insulation containing asbestos and mineral fibres from components in the controlled area of the plant.

To check compliance with radiological 'free release' limits, the material removed needed to be inventoried and its contamination level determined. One of the relevant radionuclides was tritium which has a very low release level

FIG. II–1. Shipping of fresh fuel assemblies from the Trino NPP.

of 0.1 Bq/g. In order to clear the materials, a suitable facility had to be constructed to measure these low activities of tritium with a low minimum detection activity (MDA) level. The major problems in measuring tritium were:

— Water separation from the sample matrix,
— Condensation of moisture,
— Measurement of activity in condensed moisture.

II–2.2. Solution found

The problem was solved by designing and building a facility (Fig. II–2) that could separate moisture from the sample using warm air. The moisture was then condensed and measured using liquid scintillation techniques. The system was calibrated by taking into account the efficiency of moisture separation, the background tritium levels, air speed and air quantity, to optimize moisture separation. Using this technique, tritium is detectable at minimum levels less than the free release levels. To obtain a good MDA level, experience shows that a sample amount of 50 L is needed; the MDA was 5.4×10^{-5} Bq/g.

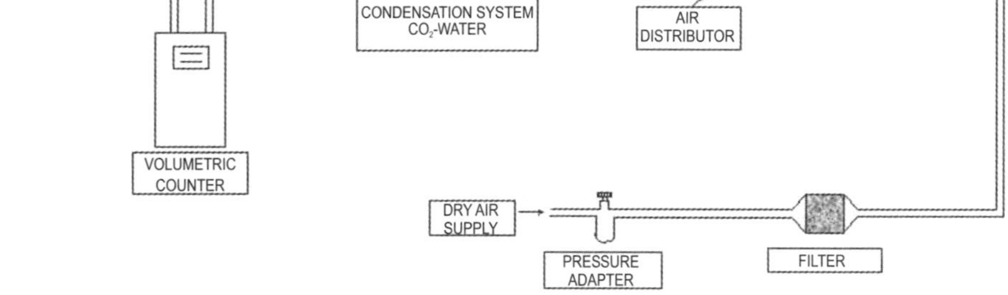

FIG. II–2. The thermal insulation moisture separator system for tritium measurements at the Trino NPP.

II–2.3. Lesson learned

During the transition period, problems may arise during the subsequent dismantling phase that may require purpose built equipment to be produced that may have a long design and manufacturing lead time. This necessitates collaboration between the NPP departments.

II–3. JASON (UK) — CASE No. 1

II–3.1. Problem

During the decommissioning of Jason it was found that the operational, maintenance, modification and repair records produced following its installation in 1962 were very comprehensive and had been scrupulously kept up to date by dedicated operating staff. However, the original reactor construction and installation records (dating from about 1960) and the records of previous nuclear/radiation related operations in the reactor hall prior to the installation of Jason were not so comprehensive. The lack of the original installation records had the potential to delay completion of the decommissioning project, particularly during the final stages when unexpected tritium contamination was discovered that caused difficulty in meeting the site radiological clearance criterion.

II–3.2. Solution found

As work progressed, the decommissioning contractors made good use of the resident JDS's knowledge of the reactor layout and its previous operational history, noting that he had been Jason's operational manager prior to the transition period. In addition, the whereabouts of retired operational personnel became increasingly important as they were able to provide the project with valuable insights and suggestions as the work progressed.

This first hand knowledge became crucial during the final stages of decommissioning when unexpected extensive tritium contamination was found in the concrete floors just outside the reactor hall. This led directly to a two month delay in completing the project. This tritium contamination was caused by previous neutron accelerator operations that predated Jason operations. These previous operations came to light primarily through personal contacts with retired personnel rather than through existing operating records. Had this matter been known about or considered earlier, the project delay might have been avoided.

II–3.3. Lesson learned

The site licensee should ensure that comprehensive and accurate records are kept of nuclear facility construction, installation, operation, modification and previous building use, and made fully available at the start of any decommissioning project. In the absence of detailed records, the site licensee should ensure that a member of his resident oversight staff has long-standing previous operational experience of facility operations and, ideally, knowledge of the whereabouts of previous operational employees.

II–4. JASON (UK) — CASE No. 2

II–4.1. Problem

Early on in the Jason decommissioning project it was realized that a formal protocol would be required to define the strategy for the final radiological survey of the site and to determine the actual radiological site release criteria. In addition to the necessary site cleanup and internal survey, it was recognized that the final IRS should include the whole site, including the decommissioning facility. It was also considered that this survey should be carried out by an organization that had a high international and national standing in the radiological field.

II–4.2. Solution found

During the transition period numerous meetings were held between the NII, the EA, the Naval Regulatory Authority and the site licensee to determine the necessary radiological survey and clearance criteria protocols. These were finally agreed about midway through the decommissioning project. The NRPB was contracted to carry out the IRS to the agreed protocol and provide the EA with a comprehensive survey report for each area surveyed. From a technical and public relations point of view, the NRPB was considered to be one of the most suitable, independent, competent and widely recognized organizations available in the UK to undertake this final radiological survey.

The agreed radiological clearance criteria for the reactor areas were:

(a) The general criteria for radiological clearance of the site to unrestricted release will be the removal of all identified contamination from the operation of the Jason reactor;

(b) Levels of artificial contamination remaining in the environment following removal of the reactor should not on average exceed the 'substances of low level exemption order' of 0.4 Bq/g;
(c) Any residual levels of external radiation must be compared with natural background environmental measurements from other parts of Greenwich and shown to be equivalent, taking into account local natural variation.

The criteria for radiological clearance of the other buildings on the college site were that:

(1) No loose contamination remains on the surfaces (including drains);
(2) Fixed alpha contamination levels do not exceed 0.04 Bq/cm^2 (0.1 DL_{α});
(3) Fixed beta/gamma contamination levels do not exceed 0.4 Bq/cm^2 (0.1 $DL_{\beta/\gamma}$);
(4) Any residual levels of external radiation must not exceed natural background environmental measurements from other parts of Greenwich.

The agreed protocol stated that the EA would issue revocation notices for the accumulation, storage, discharge and disposal of radioactive substances on the site once they were satisfied that the site had been comprehensively surveyed and met their release criteria. This would signify that the site could subsequently be used without any future radiological restriction.

II–4.3. Lesson learned

The site licensee should ensure that a comprehensive and early radiological survey and site release criteria protocol is worked up and agreed to by the relevant regulatory authorities and environmental protection agencies early on in the decommissioning project. He should also ensure that the final IRS of the complete site, as applicable, is carried out by the most suitable, independent and competent organization available in order to satisfy both public scrutiny and high levels of technical credibility.

II–5. CAORSO NPP, ITALY

II–5.1. Problem

Four heat exchangers which were located high up in the turbine hall had to be removed. Each had an outer diameter of 1500 mm, a length of 9200 mm, and weighed about 22 000 kg. The internals comprised 1770 cupronickel pipes. They were located in pairs on concrete foundations (Fig. II–3).

To dismantle these, the project had to take account of their location, their size and other constraints, including adjacent equipment. It was planned to remove the heat exchangers by attaching suitable lifting equipment so that it could be moved by using air cushions and subsequently the turbine hall bridge crane to an area where the final dismantling could be carried out using an oxyacetylene torch.

While planning the job, insufficient consideration was given to the stability of each pair of heat exchangers and the lack of access required to manoeuvre the lifting equipment. Additionally, it was not realized that movement over an irregular and uneven floor would be difficult.

II–5.2. Solution found

Upon reconsideration of the problem, it was judged to be safer and more convenient to dismantle the heat exchangers into segments that could be disposed of without further processing by cutting them in situ using an oxygen cutter. Thus a segmenting technique suitable for pipe systems was tried out and demonstrated. The operation was carried out not only safely but with lower costs than would have been incurred using the original methods.

FIG. II–3. Caorso NPP heat exchanger.

II–5.3. Lesson learned

It is important when planning and designing such tasks that field operators be involved and apply their experience to the plans.

II–6. WÜRGASSEN NPP, GERMANY

II–6.1. Problem

During the post-operational phase of an NPP, systems and components are drained and flushed. Contaminated liquids have to be moved from the circuits in order to drain certain parts of the systems, e.g. for gaining access for sampling, restructuring, fitting of primary circuit decontamination equipment, etc. During these activities it is often necessary to move contaminated liquids along different routes than during operation entailing, for example, temporary connections between parts of the systems which have not been in direct contact before, reversed circulation, use of pipes for contaminated liquids which formerly carried different media, e.g. steam. In this way the contamination profile within the plant is significantly changed and will not correspond to what would be expected during normal operation. Contamination will be transferred to parts of the system that previously possessed a totally different nuclide composition (vector).

At the Würgassen NPP in Germany, the system decontamination and other activities during the transition period have led to significant changes in nuclide vectors in some parts of the systems. Assigning nuclide vectors to material in the course of clearance procedures has thus been complicated, plant specific clearance regulations have become more complex, and clearance of materials has been delayed.

II–6.2. Solution found

In order to determine the appropriate nuclide vectors for the Würgassen NPP the sampling programme had to be substantially increased. Significant variations of activity percentages for certain nuclides were even observed between adjacent samples. The results of this sampling programme were analysed in the light of special operation modes during outages and during the transition period (especially system decontamination). In this way unexpected nuclide vectors could be explained.

II–6.3. Lessons learned

In order to avoid cross-contamination between systems and contamination of initially uncontaminated parts of components, careful planning and documentation for draining and flushing during the transition period is recommended. Full documentation of all operating modes during normal operation, outages and the transition period is essential during decommissioning, provided that the proper information is transferred and used. In addition, careful planning of the sampling programme, in particular of the number of samples which have to be taken, is essential to ensure a good radiological characterization of the plant for decontamination planning and material clearance.

II–7. KOREAN RESEARCH REACTORS — CASE No. 1

II–7.1. Problem

During the transition period, the source term estimation was carried out on activated areas such as the bio-shield concrete, the RSR, the reactor pool liner, the shielding plate and the graphite block inside the thermal column. However, the activity could not be estimated correctly because the real composition of the impurities in the activated material could not be ascertained from the design documents. For example, the amount of ^{59}Fe in aluminium significantly affected the total activity of activated material.

II–7.2. Solution found

The source terms were estimated from the composition of activated material in a similar reactor (the TRIGA Mark-II ICI reactor, UK). This reactor type is similar to KRR-1 in its maximum thermal power.

II–7.3. Lesson learned

To estimate source terms correctly it is necessary to know their actual composition, including impurities of the activated materials. Therefore, the physical and chemical properties of materials should be recorded in the original design documents.

II–8. KOREAN RESEARCH REACTORS — CASE No. 2

II–8.1. Problem

In the Republic of Korea, the cost of decommissioning commercial NPPs will be met from a decommissioning fund established by the Korea Hydraulic and Nuclear Power Company based on a kW·h levy. However, this was not the case for the research reactors. Consequently, planning for the decommissioning of KRR-1 and KRR-2 was delayed.

II–8.2. Solution found

Three years after shutdown of KRR-1 and KRR-2, the decommissioning was funded from the government budget.

II–8.3. Lesson learned

Funds for the decommissioning of all nuclear installations, including research reactors, should be allocated during the operational period as is currently the case for commercial NPPs.

II–9. A-1 REACTOR, SLOVAKIA

II–9.1. Problem

A series of problems was encountered during the transition period of the A-1 reactor which had been shut down in 1977 following an accident. Common features of these problems were the lack of documentation, appropriate methodologies, infrastructure, procedures and experience in decommissioning. This situation arose from a combination of the legal framework in existence at that time and from the fact that no preparatory activities for decommissioning were carried out during the operating period. Decommissioning of the A-1 reactor could only proceed when the following issues had been addressed:

(a) Preparation of the necessary documentation;
(b) Development of the appropriate methodologies and procedures;
(c) Establishment of the legislative framework for decommissioning;
(d) Development and procurement of equipment for treatment and conditioning of radioactive waste;

(e) Development and construction of a repository for conditioned radioactive waste.

II–9.2. Solution found

Based on the experience of decommissioning the A-1 reactor, detailed technical planning and preparation of relevant documentation for the shutdown of the V-1 NPP will enable the transition from operation to decommissioning to proceed in accordance with an optimized schedule.

II–9.3. Lesson learned

In order to avoid uncertainties and delays during the transition period it is important to plan the decommissioning and prepare the documentation in a timely manner.

DEFINITIONS

The definitions given below are taken from the IAEA Safety Glossary, Rev. April 2000 except where marked with an asterisk (*).

decommissioning. Administrative and technical actions taken to allow the removal of some or all of the regulatory controls from a facility (except for a repository which is closed and not decommissioned).

decommissioning plan. A document containing detailed information on the proposed decommissioning of a facility.

nuclear facility. A facility and its associated land, buildings and equipment in which radioactive materials are produced, processed, used, handled, stored or disposed of on such a scale that safety must be considered.

operation. All activities performed to achieve the purpose for which a facility was constructed.

operational period (operating period). The period through which a facility is being used for its intended purpose until decommissioning or closure.

permanent shutdown*. That point in the life of a plant when it reaches the end of the operational period, at which point there is no intent to restart operations.

safe enclosure (during decommissioning)*. A condition of a nuclear facility during the decommissioning process in which only surveillance and maintenance of the facility takes place.

transition period*. The period through which the administrative and technical activities to take the plant from the operational period to placement in a safe, stable and known condition in preparation for SE and/or dismantling are planned and implemented.

CONTRIBUTORS TO DRAFTING AND REVIEW

Blažek, J.	Slovak Electric, Slovakia
Coles, S.	United Kingdom Atomic Energy Authority, United Kingdom
Daniška, V.	Decom, Slovakia
Del Lucchese, M.	Societa' Gestione Impianti Nucleari, Italy
Ghioni, G.	Societa' Gestione Impianti Nucleari, Italy
Hamblin, C.	United Kingdom Atomic Energy Authority, United Kingdom
Hutchison, S.	Health and Safety Executive, United Kingdom
Johansson, L.	Barsebäck Nuclear Power Plant, Sweden
Laraia, M.	International Atomic Energy Agency
Lee, K.-W.	Korea Atomic Energy Research Institute, Republic of Korea
Lockwood, R.J.S.	HMS Sultan, United Kingdom
Lorentz, H.	Barsebäck Nuclear Power Plant, Sweden
Madrid, F.	Empresa Nacional de Residuos Radioactivos, Spain
Meyer, F.	Austrian Research Centers, Seibersdorf, Austria
Mittler, M.	Hamburgische Electrizitäts-Werke AG, Germany
Negin, C.	Project Enhancement Corp., United States of America
Nokhamzon, J.-G.	Commissariat à l'énergie atomique, France
Ølgaard, P.	Risø National Laboratory, Denmark
Podlaha, J.	Nuclear Research Institute Rež, Czech Republic
Reisenweaver, D.	International Atomic Energy Agency
Thierfeldt, S.	Brenk Systemplanung GmbH, Germany
Václav, J.	Nuclear Regulatory Authority, Slovakia
Van der Hulst, P.	Joint Dutch Nuclear Power Plant, Netherlands
Vignau, B.	CODEM, France

Consultants Meetings

Vienna, Austria: 19–23 February 2001, 3–7 February 2003

Technical Committee Meeting

Vienna, Austria: 4–8 March 2002

No. 17, January 2003

WHERE TO ORDER IAEA PUBLICATIONS

In the following countries IAEA publications may be purchased from the sources listed below, or from major local booksellers. Payment may be made in local currency or with UNESCO coupons.

AUSTRALIA Hunter Publications, 58A Gipps Street, Collingwood, Victoria 3066
Telephone: +61 3 9417 5361 • Fax: +61 3 9419 7154
E-mail: admin@tekimaging.com.au • Web site: http://www.hunter-pubs.com.au

BELGIUM Jean de Lannoy, avenue du Roi 202, B-1190 Brussels • Telephone: +32 2 538 43 08 • Fax: +32 2 538 08 41
E-mail: jean.de.lannoy@infoboard.be • Web site: http://www.jean-de-lannoy.be

CANADA Renouf Publishing Company Ltd., 1-5369 Canotek Rd., Ottawa, Ontario, K1J 9J3
Telephone: +613 745 2665 • Fax: +613 745 7660
E-mail: order.dept@renoufbooks.com • Web site: http://www.renoufbooks.com

CHINA IAEA Publications in Chinese: China Nuclear Energy Industry Corporation, Translation Section, P.O. Box 2103, Beijing

FINLAND Akateeminen Kirjakauppa, PL 128 (Keskuskatu 1), FIN-00101 Helsinki
Telephone: +358 9 121 4418 • Fax: +358 9 121 4435
E-mail: sps@akateeminen.com • Web site: http://www.akateeminen.com

FRANCE Nucléon Editions, 20, rue Rosenwald, F-75015 Paris
Telephone: +33 1 42 50 15 50 • +33 6 86 92 69 39 • Fax +33 1 53 68 14 93 • *E-mail: nucleon@nucleon.fr*
Form-Edit, 5, rue Janssen, P.O. Box 25, F-75921 Paris Cedex 19
Telephone: +33 1 42 01 49 49 • Fax: +33 1 42 01 90 90 • *E-mail: formedit@formedit.fr*

GERMANY UNO-Verlag, Vertriebs- und Verlags GmbH, Am Hofgarten 10, D-53113 Bonn
Telephone: +49 228 94 90 20 • Fax: +49 228 94 90 222
E-mail: bestellung@uno-verlag.de • Web site: http://www.uno-verlag.de

HUNGARY Librotrade Ltd., Book Import, P.O. Box 126, H-1656 Budapest
Telephone: +36 1 257 7777 • Fax: +36 1 257 7472 • *E-mail: books@librotrade.hu*

INDIA Allied Publishers Limited, 1-13/14, Asaf Ali Road, New Delhi 110002
Telephone: +91 11 3233002, 004 • Fax: +91 11 3235967
E-mail: aplnd@del2.vsnl.net.in • Web site: http://www.alliedpublishers.com

ITALY Libreria Scientifica Dott. Lucio di Biasio "AEIOU", Via Coronelli 6, I-20146 Milan
Telephone: +39 02 48 95 45 52 or 48 95 45 62 • Fax: +39 02 48 95 45 48

JAPAN Maruzen Company, Ltd., 13-6 Nihonbashi, 3 chome, Chuo-ku, Tokyo 103-0027
Telephone: +81 3 3275 8582 • Fax: +81 3 3275 9072
E-mail: journal@maruzen.co.jp • Web site: http://www.maruzen.co.jp

NETHERLANDS Martinus Nijhoff International, Koraalrood 50, P.O. Box 1853, 2700 CZ Zoetermeer
Telephone: +31 793 684 400 • Fax: +31 793 615 698 • *E-mail: info@nijhoff.nl • Web site: http://www.nijhoff.nl*
Swets and Zeitlinger b.v., P.O. Box 830, 2160 SZ Lisse
Telephone: +31 252 435 111 • Fax: +31 252 415 888 • *E-mail: infoho@swets.nl • Web site: http://www.swets.nl*

SLOVAKIA Alfa Press, s.r.o, Račianska 20, SQ-832 10 Bratislava • Telephone/Fax: +421 7 566 0489

SLOVENIA Cankarjeva Zalozba d.d., Kopitarjeva 2, SI-1512 Ljubljana
Telephone: +386 1 432 31 44 • Fax: +386 1 230 14 35
E-mail: import.books@cankarjeva-z.si • Web site: http://www.cankarjeva-z.si/uvoz

SPAIN Díaz de Santos, S.A., c/ Juan Bravo, 3A, E-28006 Madrid
Telephone: +34 91 781 94 80 • Fax: +34 91 575 55 63 • *E-mail: compras@diazdesantos.es • carmela@diazdesantos.es barcelona@diazdesantos.es • julio@diazdesantos.es • Web site: http://www.diazdesantos.es*

UNITED KINGDOM The Stationery Office Ltd, International Sales Agency, 51 Nine Elms Lane, London SW8 5DR
Telephone: +44 870 600 552 • Fax: +44 207 873 8416
E-mail: Orders to: book.orders@theso.co.uk • Enquiries to: ipa.enquiries@theso.co.uk
Web site: http://www.the-stationery-office.co.uk

On-line orders

DELTA Int. Book Wholesalers Ltd., 39 Alexandra Road, Addlestone, Surrey, KT15 2PQ
E-mail: info@profbooks.com • Web site: http://www.profbooks.com

Books on the Environment
SMI (Distribution Services) Limited, P.O. Box 119, Stevenage SG1 4TP, Hertfordshire
E-mail: customerservices@earthprint.co.uk • Web site: http://www.earthprint.co.uk

UNITED STATES OF AMERICA Bernan Associates, 4611-F Assembly Drive, Lanham, MD 20706-4391
Telephone: 1-800-274-4447 (toll-free) • Fax: (301) 459-0056/1-800-865-3450 (toll-free)
E-mail: query@bernan.com • Web site: http://www.bernan.com
Renouf Publishing Company Ltd., 812 Proctor Ave., Ogdensburg, NY, 13669
Telephone: +888 551 7470 (toll-free) • Fax: +888 568 8546 (toll-free)
E-mail: order.dept@renoufbooks.com • Web site: http://www.renoufbooks.com

Orders and requests for information may also be addressed directly to:

Sales and Promotion Unit, International Atomic Energy Agency
Wagramer Strasse 5, P.O. Box 100, A-1400 Vienna, Austria
Telephone: +43 1 2600 22529 (or 22530) • Fax: +43 1 2600 29302
E-mail: sales.publications@iaea.org • Web site: http://www.iaea.org/worldatom/Books

04-00501

A. F. JOFFÉ

Halbleiter-Thermoelemente

A. F. JOFFÉ

Halbleiter-Thermoelemente

Herausgeber der deutschen Ausgabe
Dr. Ernst A. Niekisch

AKADEMIE-VERLAG · BERLIN

1957

A. Ф. Иоффе
Полупроводниковые термоэлементы

Erschienen im Verlag der Akademie der Wissenschaften der UdSSR
Moskau-Leningrad 1956

Übersetzt aus dem Russischen von Dipl.-Phys. H. Vogel

Erschienen im Akademie-Verlag GmbH., Berlin W 8, Mohrenstraße 39
Lizenz-Nr. 202 · 100/481/57

Satz und Druck: Druckhaus „Maxim Gorki“, Altenburg
Bestell- und Verlagsnummer: 52 67
Printed in Germany

INHALT

VORWORT

Mehr als 130 Jahre sind seit der Entdeckung der thermoelektrischen Erscheinungen vergangen. Verhältnismäßig früh erkannte man, daß sie sich für die direkte Umwandlung von Wärme in elektrische Energie eignen. Bis in allerjüngste Zeit wurden sie aber nur für meßtechnische Zwecke und zum Nachweis von Strahlungsenergie benutzt. Versuche, mit ihrer Hilfe Akkumulatoren — als Ersatz für galvanische Elemente — aufzuladen, blieben in den Anfängen stecken.

Es wurde eine ganze Reihe von Verfahren zur Umwandlung der Strahlungsenergie der Sonne in elektrische Energie mit Hilfe von Thermoelementen vorgeschlagen. Auch die Möglichkeit einer thermoelektrischen Kühlung wurde eingehend erörtert. Praktische Lösungen jedoch wurden hier nicht gefunden, da Wissenschaftler und Techniker ihre ganze Aufmerksamkeit auf die Metalle, die einzigen in der damaligen Elektrotechnik verwendeten Leiter, richteten. Obwohl SEEBECK schon in seiner ersten klassischen Untersuchung aus dem Jahre 1822 neben den Metallen auch Mineralien und Erze untersuchte und die Vorzüge der Halbleiter — nämlich des Bleisulfids und einer Zink-Antimon-Legierung — gegenüber den Metallen bemerkte, obwohl die hohen Werte der Thermospannungen von Halbleitern schon seit 50 Jahren bekannt sind, wurden sie in Thermoelementen nicht ausgenutzt.

Wie die amerikanische Ingenieurzeitschrift „Electronics" mitteilt, verwendete während des 2. Weltkriegs die US-Armee metallische Thermoelemente mit einem Wirkungsgrad von etwa 0,2%. Bei der Abschätzung der Zukunftsaussichten des Thermoelements als Quelle elektrischer Energie kommt die Zeitschrift zu dem Schluß, daß man den Wirkungsgrad von Thermoelementen nur auf 8% erhöhen müßte, um eine vollständige Revolution in der Technik kleiner Energien (d. h. der Technik der Generatoren mit einer Leistung unterhalb 100 kW) hervorzurufen.

In den letzten Jahren sind außerhalb der Sowjetunion etwa 50 Arbeiten über Thermoelemente sowohl hinsichtlich ihrer Anwendung zur Erzeugung von elektrischer Energie als auch zur Kälteerzeugung erschienen. Theoretisch und praktisch wesentliche Ergebnisse enthalten aber alle diese Arbeiten nicht.

In der Sowjetunion wird zwar die Forschungsarbeit über thermoelektrische Probleme forciert — es werden beispielsweise thermoelektrische Generatoren zum Betrieb von Radioempfängern entwickelt und produziert, die die Abwärme von Petroleumlampen ausnutzen —, aber ihrem Maßstabe nach entsprechen diese Arbeiten offensichtlich nicht der Wichtigkeit des Problems. Es ist an der Zeit, die Entwicklung auf dem sich bereits klar abzeichnenden Weg weiter zu fördern und die Thermoelemente über den Bereich des Laboratoriums hinauszubringen. Die Lösung dieses Problems erfordert die vereinten Anstrengungen von Physikern und Chemikern, Wärmetechnikern und Elektroingenieuren; es ist auch im gegenwärtigen Stadium hervorragend geeignet, das Interesse unserer technisch am weitesten entwickelten Betriebe zu wecken.

Hierfür ist aber vor allem erforderlich, daß die grundlegenden Kenntnisse über Halbleiter und speziell über Halbleiter-Thermoelemente einem weiten Kreise nahegebracht werden. Man sollte dem Chemiker und dem Ingenieur das notwendige Wissen und die bereits gesammelten Erfahrungen übermitteln, um seine schöpferische Initiative anzuregen.

Dem erstgenannten Ziel, nämlich einer Einführung in die Physik der Halbleiter, dient eine Schriftenreihe des Verlages der Akademie der Wissenschaften der UdSSR: Mein Buch „Die Halbleiter in der modernen Physik“ (1954) und eine Reihe von Monographien über Einzelgebiete, die bibliographische Arbeit von W. P. SHUSE „Wissenschaftliche Literatur über Halbleiter“ (1955), die mehr als 8000 Titel umfaßt, und schließlich mein populärwissenschaftliches Büchlein „Halbleiter“ (1955). Die Zeitschrift „Успехи Физических Наук“ (Fortschr. physik. Wiss.) hat im Jahre 1953 (Band 49, Heft 2/3) einen Bericht von A. G. SAMOILOWITSCH und L. I. KORENBLIT veröffentlicht[1]), der sich mit der Theorie der thermoelektrischen Erscheinungen beschäftigt.

Dieses Buch und die fast gleichzeitig erschienene Kollektivarbeit von A. F. JOFFÉ, L. S. STILBANS, T. S. STAWITZKI und J. K. JORDANISCHWILI über „Thermoelektrische Kühlung“ haben zum Ziel, unsere Erfahrungen auf diesem Gebiet zusammenzufassen und dem Wissenschaftler und Techniker die notwendigen Kenntnisse zu vermitteln.

Der Schwerpunkt dieses Buches liegt im zweiten Kapitel. Das erste Kapitel (Einleitung) und das dritte (ergänzende) Kapitel fassen entweder — um dem Leser das Verständnis zu erleichtern — kurz die Tatsachen zusammen, die in der Literatur bereits veröffentlicht wurden, oder sie behandeln neue noch nicht ausgenutzte Möglichkeiten, die zu einer richtigen Einschätzung des Hauptproblems, nämlich des thermoelektrischen Generators aus Halbleitermaterialien, verhelfen sollen.

Ich möchte an dieser Stelle für die Hilfe danken, die mir von L. S. STILBANS zukam; er berichtigte eine Reihe von Fehlern und Unklarheiten; ferner danke ich A. I. BURSTEIN, Student an der Staatlichen Universität Odessa und B. I. BOK, Student an dem Polytechnischen Institut Leningrad für die Nachprüfung der Rechnungen.

[1]) s. Fußnote auf S. 19.

KAPITEL

Einleitung

I. Die Entdeckung der thermoelektrischen Erscheinungen

In den Jahren 1822/23 beschrieb T. J. SEEBECK in den „Abhandlungen der königlichen Akademie der Wissenschaften (Berlin)“ eine Erscheinung, die er als ein Auftreten von freiem Magnetismus in miteinander verbundenen verschiedenartigen Leitern unter dem Einfluß einer Temperaturdifferenz deutete. Aus der Beschreibung seiner Experimente geht hervor, daß SEEBECK die thermoelektrischen Ströme entdeckt hatte, die in einem geschlossenen Kreis aus verschiedenartigen Leitern auftreten, wenn die Kontaktstellen auf verschiedenen Temperaturen gehalten werden.

Diese Entdeckung erfolgte in einer Epoche, als auf die Entdeckungen OERSTEDs über die Einwirkung eines elektrischen Stroms auf eine Magnetnadel eine ganze Reihe von Arbeiten von AMPÉRE, BIOT, SAVART, LAPLACE u. a. folgte, in denen die Wechselwirkung zwischen elektrischen Strömen und Magnetfeldern aufgeklärt wurde; die Hypothese der AMPÉREschen Molekularströme führte die magnetischen Erscheinungen auf elektrische Ströme zurück. Von einem Montag zum anderen (am Montag fanden die Sitzungen der Pariser Akademie der Wissenschaften statt) wuchs die Kenntnis über die elektromagnetischen Erscheinungen. Physiker, Mathematiker und Chemiker beteiligten sich an der Entwicklung dieses neuen Wissensgebietes.

Von Paris aus dehnte sich diese Forschungstätigkeit weiter aus: zunächst nach der Schweiz und nach Deutschland, dann auch nach England, wo FARADAY die Kenntnisse der Physiker durch die Entdeckung der elektromagnetischen Induktion bereicherte.

Betrachtet man die Versuche SEEBECKs unvoreingenommen, so kann man kaum daran zweifeln, daß die von ihm beobachteten Erscheinungen auf dem elektrischen Strom beruhen (Verschwinden der magnetischen Wirkungen, sobald der Strom durch Einfügen einer nichtleitenden Schicht unterbrochen wurde, quantitativer Zusammenhang zwischen dem Magnetfeld eines Stromes und den beobachteten Abweichungen der Magnetnadel usw.). SEEBECK verzichtete jedoch nicht nur auf diese naheliegende Deutung seiner Entdeckung, sondern setzte sich sogar auch aktiv noch mehrere Jahre gegen sie ein; er warf den Vertretern der Theorie der thermoelektrischen Ströme vor, sie wären von einer „Modekrankheit“ ergriffen, die durch die Entdeckungen OERSTEDs ausgelöst worden sei.

Den Grund für eine so voreingenommene Haltung SEEBECKs gegenüber seiner eigenen Entdeckung findet man im Schlußteil seiner umfangreichen Abhandlung, wo er versucht, den Erdmagnetismus mit der Temperaturdifferenz zwischen dem Äquator oder einer Anzahl südlicher Vulkane und den polaren Eisgebieten zu verknüpfen. Offenbar lag ihm diese Hypothese näher am Herzen als seine Entdeckung einer weiteren Quelle des elektrischen Stroms.

Der Wert der SEEBECKschen Arbeit liegt natürlich nicht in dieser Hypothese (hatte er doch selbst gezeigt, daß eine Temperaturdiffernz ohne einen geschlossenen elektrischen Stromkreis keine magnetischen Erscheinungen hervorruft), auch nicht in dem Versuch, den elektrischen Strom zu leugnen, sondern in dem gewaltigen experimentellen Material, das SEEBECK angehäuft hat und das die verschiedenartigsten festen und flüssigen Metalle, ihre Legierungen sowie Mineralien und Halbleiter umfaßt. SEEBECKs Irrtum spielte eine durchaus förderliche Rolle: um die elektrische Herkunft der thermoelektrischen Ströme zu widerlegen, verglich er bei den verschiedensten Materialien die Elektrisierung (das Kontaktpotential) bzw. die VOLTAsche Spannungsreihe mit der Einwirkung einer Temperaturdifferenz auf eine Magnetnadel und wies die Unterschiede nach.

Die von SEEBECK aufgestellte umfangreiche thermoelektrische Reihe ist auch heute noch von Interesse. In unseren Bezeichnungen (α: Thermo-EMK pro Grad[1]) und σ: spezifische elektrische Leitfähigkeit) wird die SEEBECKsche Reihe durch das Produkt $\alpha\sigma$ gekennzeichnet; da, wie wir heute wissen, auch die spezifische Wärmeleitfähigkeit $\varkappa$ maßgebend für die thermoelektrischen Eigenschaften eines Materials ist, ziehen wir die Größe $\alpha^2\sigma/\varkappa$ vor.

Die Sorgfalt der Untersuchungen SEEBECKs wird beispielsweise dadurch veranschaulicht, daß in einer Arbeit von Maria TELKES, die in einer amerikanischen Zeitschrift 125 Jahre später veröffentlicht wurde, als günstigstes Paar die Materialien ZnSb und PbS angegeben werden, die auch in der SEEBECKschen Reihe ganz außen stehen.

Ein Vergleich der qualitativen Reihe von SEEBECK mit modernen Messungen der Thermo-EMK von Metallen zeigt, wie gut die Reihe stimmt. Die Unterschiede der SEEBECKschen Reihe gegenüber den Reihen von JUSTI und MEISSNER sind nicht größer als die Unterschiede dieser beiden Reihen untereinander. Nachstehend geben wir die drei thermoelektrischen Reihen an (in den Reihen von JUSTI und MEISSNER sind die Werte der Thermo-EMK in μV/Grad angegeben).

Hätte SEEBECK gleich nach seiner Entdeckung versucht, das Thermoelement zur Erzeugung elektrischer Energie zu verwenden, so hätte er mit Hilfe der äußersten Glieder seiner Reihe Wirkungsgrade von etwa 3% erreichen können, also Werte, wie sie die besten damaligen Dampfmaschinen lieferten.

Wir müssen noch anführen, daß SEEBECK auch schon thermoelektrische Erscheinungen bemerkte, die durch eine Temperaturdifferenz innerhalb eines homogenen Materials hervorgerufen werden, und daß er die auftretenden Kräfte qualitativ abschätzte; W. THOMSON entdeckte diese 30 Jahre später nochmals und bewies

[1]) Für genaue Definition und Nomenklatur s. Fußnote auf S. 7.

mittels einer thermodynamischen Behandlung der thermoelektrischen Prozesse die Existenz solcher thermoelektrischen Kräfte.

SEEBECK 1822	JUSTI 1948		MEISSNER 1955			
			Metalle		Halbleiter	
PbS	Bi	−80	Bi	−70	MoS	−770
Bi	Co	−21	Co	−17.5	ZnO	−714
Bi-Amalgam	Ni	−20	Ni	−18	CuO	−696
Ni	K	−14	K	−12	F_2O_3 (400° C)	−613
Co	Pd	−8	Pd	−6	FeO	−500
Pd	Na	−7	Na	−4.4	Fe_3O_4	−430
Pt Nr 1	Pt	−5	Pt	−3.3	FeS_2	−200
U	Hg	−5	Hg	−3.4	MgO_3H_2	−200
Au Nr 1	C	−3.5			SnO	−139
Cu Nr 1	Al	−1.5	Al	−0.6	Fe_2O_3 (50°C)	−60
Rh	Mg	−1.5	Mg	−0.4	CdO	−41
Au Nr.2	Pb	−1.0	Pb	−0.1	CuS	−7
Ag	Sn	−1.0	Sn	+0.1	FeS	+26
Zn	Cs	−0.5	Cs	+0.2	CdO	+30
C	Y	−1.0	Y	+2.2	$GeTiO_3$	+140
Cu Nr 3	Rh	+1.0	Rh	+2.5	NiO	+240
Pt Nr 4	Zn	+1.5	Zn	+2.9	Mn_2O_3	+385
Cd	Ag	+1.5	Ag	+2.4	Cu_2O	+474
Stahl	Au	+1.5	Au	+2.7	Cu_2O	+1000
Fe	Cu	+2.0	Cu	+2.6	Cu_2O	+1120
As	W	+2.5	W	+1.5	Cu_2O	+1150
Sb	Cd	+3.5	Cd	+2.8		
SbZn	Mo	+6.5	Mo	+5.9		
Fe	Fe	+12.5	Fe	+16		
	Sb	+42	Sb	+35		
	Si	+44				
	Fe	+49	Fe	+400		
	Se		Se	+1000		

12 Jahre nach SEEBECKs Entdeckung veröffentlichte der Uhrmacher PELTIER in den französischen „Annales de physique et chimie" im Jahre 1834 eine Arbeit über die Temperaturanomalien, die er in der Nähe der Grenzfläche zweier verschiedener Leiter beim Durchgang eines elektrischen Stroms beobachtet hatte. Die Erscheinung, die PELTIER als erster beobachtete und die daher PELTIER-Effekt heißt, besteht in der Erzeugung oder dem Verbrauch einer Wärmemenge Q (je nach der Stromrichtung) an der Grenze zweier verschiedener Leiter beim Durchgang eines Stroms I; es gilt

$$Q = \Pi I,$$

wobei Π der PELTIER-Koeffizient ist.

Der PELTIER-Effekt hängt eng mit der Thermoelektrizität zusammen; zwischen dem Koeffizienten Π und der Thermokraft α[1]) besteht folgende Beziehung:

$$\Pi = \alpha T,$$

[1]) Vgl. Fußnote 2 auf S. 9.

wobei T die absolute Temperatur des Kontaktes ist. Eine Temperaturdifferenz ruft in einem geschlossenen Kreis aus verschiedenen Materialien einen elektrischen Strom hervor; umgekehrt erzeugt ein Strom, der durch diesen Kreis fließt, eine Temperaturdifferenz. Eigentümlicherweise erkannte PELTIER diesen Zusammenhang nicht, obwohl er seine Versuche mit einer thermoelektrischen Kette anstellte. Die Thermoelektrizität spielte hierbei aber nur eine Hilfsrolle als Quelle schwacher Ströme.

Die von PELTIER beobachteten Anomalien äußerten sich um so stärker, je höher die Thermo-EMK war. Besonders deutlich waren sie bei dem Paar Wismut-Antimon. PELTIER suchte und glaubte in seinen Versuchen eine Bestätigung für seine vorgefaßte Meinung zu finden, das universelle Gesetz der Wärmeerzeugung durch einen Strom, das JOULEsche Gesetz, gelte nur für starke Ströme. Bei schwachen Strömen, wie sie durch ein Thermoelement hervorgerufen werden, sollten sich nach PELTIER die individuellen Eigenschaften der Metalle durchsetzen. Diese Erwartung ließ sich bei den meisten Leitern nicht bestätigen, jedoch äußerte sich die Natur des Metalls in Anomalien an den Kontaktstellen, eben den thermoelektrischen Eigenschaften des speziellen Metalls, von denen PELTIER aber nichts wußte. Er suchte die Erklärung in der Härte oder Weichheit des Metalls und in seiner elektrischen Leitfähigkeit; wenn die Tatsachen seinen Erwartungen nicht entsprachen, wie etwa im Falle des Wismuts, dessen große thermische Anomalien PELTIER auf eine große Leitfähigkeit zurückführte, bezweifelte er einfach die Richtigkeit der Messungen.

Es vergingen mehrere Jahre, in denen BECQUEREL und andere Forscher die wahre Bedeutung des PELTIER-Effekt aufklärten, bis im Jahre 1838 der Petersburger Forscher LENZ durch einen einfachen Versuch jedem Zweifel ein Ende setzte. In eine Vertiefung der Lötstelle eines aus Wismut und Antimon zusammengesetzten Stabes setzte LENZ ein Wasserröhrchen; das Wasser gefror, wenn der Strom in der einen Richtung durchging, und taute bei der anderen Stromrichtung wieder auf. Bekanntlich erfordert das Gefrieren von 1 g Wasser den Entzug von 80 cal, das Auftauen die Zufuhr der gleichen Energiemenge. Damit war klar, daß an einer Lötstelle zweier verschiedener Leiter der elektrische Strom in Abhängigkeit von seiner Richtung Wärme erzeugt oder verbraucht.

Die thermoelektrischen Erscheinungen waren nicht dazu angetan, bei den Physikern größeres Interesse zu erwecken. In der mächtigen Welle des Elektromagnetismus, die FARADAY zur Entdeckung der elektromagnetischen Induktion führte, gingen diese scheinbar unbedeutenden Effekte unter. Die Physik ging zu den allgemeinen Gesetzen der MAXWELLschen Theorie, die Technik zur Entwicklung elektrischer Maschinen über. Nach der Entdeckung SEEBECKs verstrichen 30 Jahre, bevor im Zusammenhang mit der Thermodynamik das Interesse an allen Formen der Energieumwandlung wuchs, u. a. auch an der Umwandlung von thermischer in elektrische Energie bei den Effekten SEEBECKs und PELTIERs.

Auf diesem Wege stieß auch W. THOMSON, einer der Begründer der Thermodynamik, auf die thermoelektrischen Erscheinungen. Die thermodynamische Analyse der Effekte SEEBECKs und PELTIERs führte ihn nicht nur zur Aufstellung des bereits erwähnten Zusammenhangs zwischen diesen beiden Erscheinungen, sondern auch zur Entdeckung eines dritten Effekts, des THOMSON-Effektes, nämlich der Erzeugung oder dem Verbrauch einer Wärmemenge q beim Durchgang eines Stroms I durch einen homogenen Leiter, in dem ein Temperaturgradient $\frac{\partial T}{\partial x}$ herrscht:

$$q = \tau I \frac{\partial T}{\partial x}.$$

Die Größe τ heißt THOMSON-Koeffizient; sie hängt offenbar mit α und Π zusammen.

Wie wir bereits erwähnten, wurde der THOMSON-Effekt schon von SEEBECK experimentell beobachtet und untersucht.

Im Jahre 1857 veröffentlichte THOMSON eine Theorie der thermoelektrischen Erscheinungen in anisotropen Kristallen; aber erst im 20. Jahrhundert wies BRIDGMAN nach, daß die Richtungsänderung eines Stroms in einem Kristall mit einer Erzeugung oder einem Verbrauch von Wärme verbunden ist, ebenso wie beim PELTIER-Effekt. Dieser innere PELTIER-Effekt wurde als BRIDGMAN-Effekt bezeichnet[1]).

Fast 100 Jahre nach SEEBECKs Entdeckung, in den Jahren 1920/21 berichtete BENEDICKS über zwei weitere thermoelektrische Effekte, für die nicht eine Temperaturdifferenz zwischen den Leiterenden, sondern ein Temperaturgradient maßgebend ist. Haben die Enden eines Leiters die gleiche Temperatur, befindet sich aber dazwischen ein erhitzter oder abgekühlter Abschnitt, so tritt nach BENEDICKS zwischen den Enden eine Potentialdifferenz auf, sofern der erhitzte bzw. abgekühlte Abschnitt nicht in der Mitte des Leiters, sondern näher zu einem Ende liegt. Die sorgfältige Prüfung des BENEDICKS-Effekts durch andere Forscher hat seine Existenz nicht bestätigt. Man kann annahmen, daß der BENEDICKS-Effekt, selbst wenn er vorhanden ist, verschwindend klein gegenüber dem THOMSON-Effekt ist.

Die genannten Erscheinungen erschöpfen das Gebiet der Thermoelektrizität, wenn man von den Effekten im Magnetfeld, die als thermomagnetische Effekte bezeichnet werden, absieht.

Thermoelemente werden schon seit langem zur Temperaturmessung benutzt; ihre Anwendung in der Energetik wurde dagegen erst aktuell, als die Nachfrage nach Quellen elektrischer Energie immer größer wurde.

Im Jahre 1885 untersuchte Lord RAYLEIGH den Wirkungsgrad eines thermoelektrischen Generators und berechnete ihn (allerdings nicht ganz richtig).

Den thermoelektrischen Generator zur Erzeugung elektrischer Energie behandelte ALTENKIRCH im Jahre 1909 im wesentlichen richtig. Im nächsten Jahre (1910) untersuchte derselbe Autor das technische Problem der thermoelektrischen Kühlung und Heizung. Da aber die einzigen damals in der Technik bekannten Leiter Metalle waren, erwiesen sich diese Geräte als unwirtschaftlich. Es wurden allerdings einige Thermoelektrogeneratoren angegeben (GÜLCHER, COBLENTZ u. a.), jedoch fanden sie kaum Anwendung, da ihr Wirkungsgrad unter 0,6%, manchmal sogar unter 0,1% lag. Die Thermoelektrizität fand also wieder in dem Kuriositätenkabinett der Physik Aufnahme, zusammen mit der Lumineszenz, dem Photoeffekt und der Piezoelektrizität.

[1]) Näheres hierzu auf S. 16. — Anm. d. Red.

In der Elektrotechnik (und im Zusammenhang damit auch in der Themenstellung physikalischer Untersuchungen) änderte sich die Lage wesentlich, als man die Halbleiter näher kennenlernte. Sehr bemerkenswert für die gegenwärtige Stufe der Zusammenarbeit von Physik und Technik ist der Umstand, daß alle diese Erscheinungen neuerdings die Aufmerksamkeit der Physiker auf sich ziehen und die auf dem Gebiet der Halbleiterphysik erzielten Erkenntnisse in die Produktionstechnik eingehen. Was die Thermoelektrizität betrifft, so sollen die Wege ihrer Erforschung in diesem Buch ausführlich behandelt werden. Die Erforschung der anderen Gebiete hat in der Technik zur Entwicklung zahlreicher Typen von Leuchtröhren, sowie von keramischen Piezoelektrika, von festen und Vakuum-Photoelementen geführt.

Die Zukunftsaussichten hinsichtlich der Erreichung größerer Wirkungsgrade bei thermoelektrischen Generatoren haben sich verbessert. Im Zusammenhang mit der Industrialisierung der Sowjetunion wies ich zu Beginn des 1. Fünfjahrplans, im Jahre 1929, auf die Vorzüge eines thermoelektrischen Generators aus Halbleitermaterialien hin und gab an, daß sein Wirkungsgrad 2,5 bis 4% erreichen könne und daß eine weitere wesentliche Steigerung durchaus wahrscheinlich sei.

Tatsächlich konnte J. P. MASLAKOWEZ im Jahre 1940 über ein Thermoelement aus Bleisulfid (mit Bleiüberschuß) und Schwefel berichten, das einen Wirkungsgrad von etwa 3% hatte; im Jahre 1947 veröffentlichte M. TELKES in der amerikanischen Zeitschrift „Journal of Applied Physics" eine Arbeit, in der sie behauptete, sie habe einen Wirkungsgrad von 7% erreicht. Diese Abschätzung hat sich aber offenbar nicht bestätigt, denn im Jahre 1954 überstieg der Wirkungsgrad der von TELKES ausführlich beschriebenen Thermoelemente 3,3% nicht.

Im Jahre 1953 veröffentlichte JUSTI in der westdeutschen Zeitschrift „Kältetechnik" die Berechnung einer thermoelektrischen Kühlanlage aus metallischen und Halbleitermaterialien. Der Verfasser kommt zu dem unrichtigen Ergebnis, es gebe eine obere Grenze für die thermoelektrische Kühlung, nämlich 25° C. Daß diese Behauptung nicht zutrifft, ergibt sich schon daraus, daß unsere Thermoelemente eine Temperatursenkung um mehr als 50° C erzielen.

Im Novemberheft (1954) des „British Journal of Applied Physics" wird über eine Kühlanlage berichtet, die eine Temperatursenkung um 26° bewirkt. Die Autoren dieser Arbeit, GOLDSMID und DOUGLAS, sind Mitarbeiter der englischen Firma „The General Electric Co. Ltd.".

In den theoretischen Teilen aller dieser Arbeiten finden sich Ungenauigkeiten. Die beschriebenen praktischen Erfolge sind ebenfalls unbefriedigend und zeigen keine neuen Wege zu weiterer Verbesserung.

Unseres Erachtens sind alle energetischen Anwendungen der Thermoelektrizität, nämlich der thermoelektrische Generator, die thermoelektrische Kühlung und die thermoelektrischen Heizgeräte nicht nur von großer prinzipieller Bedeutung, sondern bieten auch reale Aussichten zu ihrer praktischen Verwirklichung auf Grund unserer heutigen Kenntnisse der Halbleiter. Außerdem ergeben sich auch unübersehbare weitere Probleme, wie z. B. die Schaffung thermoelektrischer Schall- und Ultraschallgeneratoren, die von Vakuum-Thermoelementen usw.

2. *Thermodynamische Beziehungen*

Das Experiment hat die Existenz von drei thermoelektrischen Effekten bewiesen:

1. das Auftreten einer EMK dE in einem nicht geschlossenen Kreis aus zwei verschiedenartigen Leitern, zwischen deren Enden eine Temperaturdifferenz dT besteht:

$$dE = \alpha_{1-2} dT \tag{1}$$

(α_{1-2} ist die Thermo-EMK pro Grad oder die Thermokraft zwischen den gegebenen Leitern[1]); α kann von der Temperatur abhängen;

2. die Erzeugung einer Wärmemenge dQ [Watt] an der Grenze zwischen zwei Leitern beim Durchgang eines Stroms von dI [Amp.]:

$$dQ = \Pi_{1-2} dI \tag{2}$$

(bei einer Vorzeichenänderung des Stroms ändert sich auch das Vorzeichen der erzeugten Wärmemenge);

3. Ein Strom dI erzeugt in einem Leiterabschnitt von der Länge dx eine Wärmemenge dq, wenn im Leiter ein Temperaturgradient $\frac{\partial T}{\partial x}$ besteht:

$$dq = \tau \, dI \frac{\partial T}{\partial x} dx . \tag{3}$$

Die Gesetze der Thermodynamik fordern einen bestimmten Zusammenhang zwischen den Größen α_{1-2}, Π_{1-2} und τ.

Wir betrachten einen geschlossenen Kreis aus zwei Leitern; die beiden Kontaktstellen befinden sich auf den Temperaturen T bzw. $T + dT$. Nach dem Energiesatz ist die algebraische Summe aller in der Zeiteinheit unter stationären Bedingungen im Leiter umgesetzten Energiemengen gleich Null.

Die PELTIER-Wärme in Watt, die pro Sekunde an einem der beiden Kontakte erzeugt wird, bezeichnen wir mit $\Pi_{1-2}\, dI$, die an dem anderen Kontakt verbrauchte Wärme mit $\Pi_{2-1}\, dI$, wobei gilt[2]):

$$\Pi_{2-1} = \Pi_{1-2} + \frac{\partial \Pi_{1-2}}{\partial T} dT.$$

Die pro Sekunde in einem der beiden Leiter erzeugte THOMSON-Wärme beträgt $\int_T^{T+dT} \tau_1 \, dI \, dT$, die im anderen Leiter verbrauchte Wärme $\int_T^{T+dT} \tau_2 \, dI \, dT$. Die im Kreis erzeugte elektrische Leistung beträgt $\alpha \, dT \, dI$.

[1]) In der deutschen Literatur ist die Nomenklatur nicht einheitlich. Mit *Thermo-EMK* werden wir die gesamte thermoelektrische Spannung zwischen zwei beliebigen Temperaturen bezeichnen; das durch (1) definierte α nennen wir *Thermokraft* oder *differentielle Thermospannung*. — Anm. d. Red.

[2]) Π_{1-2} bezieht sich auf die Temperatur T, Π_{2-1} auf $T + \Delta T$. — Anm. d. Red.

Wegen der Erhaltung der Energie in dem Kreis ist dann

$$\frac{\partial \Pi_{1-2}}{\partial T} dT\, dI + (\tau_1 - \tau_2)\, dI\, dT = \alpha_{1-2} dT\, dI$$

oder

$$\frac{\partial \Pi_{1-2}}{\partial T} + \tau_1 - \tau_2 = \alpha_{1-2}. \tag{4}$$

Eine zweite Beziehung zwischen α, Π und τ erhält man aus dem zweiten Hauptsatz der Thermodynamik, der sich jedoch nur auf reversible Prozesse anwenden läßt. Alle drei thermoelektrischen Effekte sind als reversibel anzusehen: sie ändern ihr Vorzeichen sowohl bei einer Vorzeichenänderung der Temperatur dT als auch bei einer Richtungsänderung des Stroms dI. In einem thermoelektrischen Kreis treten aber notwendig auch irreversible Erscheinungen wie Wärmeleitung und JOULEsche Wärme auf. Während man die letztere bei sehr schwachen Strömen vernachlässigen kann — wie wir noch sehen werden—, transportiert die Wärmeleitung Wärmemengen von derselben Größenordnung wie die PELTIER-Wärme und manchmal sogar noch mehr.

THOMSON nahm an, die thermoelektrischen Erscheinungen seien nicht organisch mit der Wärmeleitung oder der Erzeugung von Stromwärme verbunden. Für SEEBECK-, PELTIER- und THOMSON-Effekt sei nicht der Wert der Wärmeleitfähigkeit des gegebenen Materials und sein spezifischer Widerstand maßgebend, von denen die irreversiblen Prozesse im Thermoelement abhängen. Man könne diese drei Effekte einzeln betrachten, indem man Wärmeleitung und Widerstand als beliebig klein annehme und die irreversiblen Prozesse als sekundär und zufällig betrachte. Diese Überlegungen haben aber nur eine beschränkte Beweiskraft. ONSAGER gab später Bedingungen an, unter denen diese Einteilung in reversible und irreversible Prozesse zulässig ist. Seine Bedingungen gelten offenbar sowohl für Metalle als auch für Halbleiter. Man kann jedoch dieses Problem auch heute noch nicht als endgültig gelöst ansehen.

Wir wollen hier gleichwohl THOMSONs Darlegung folgen, also annehmen, daß sich der reversible Teil der thermoelektrischen Prozesse unabhängig von den irreversiblen betrachten läßt, und die daraus folgenden Schlüsse nachprüfen. Die Übereinstimmung unserer Schlußfolgerungen mit den Versuchsergebnissen kann als gewisse Bestätigung der zu Grunde liegenden Annahmen aufgefaßt werden. Da uns im folgenden nur die Schlußfolgerungen aus der Theorie interessieren werden, können wir diese Annahmen auch nur als Weg zu diesen Ergebnissen betrachten.

Wir wenden also den zweiten Hauptsatz der Thermodynamik auf die thermoelektrischen Erscheinungen in einem Thermoelement an, dessen Lötstellen sich auf den Temperaturen T_1 und T_0 befinden ($T_1 > T_0$). Die heiße Lötstelle nimmt eine Wärmemenge $\Pi_1 I$ auf, die kalte Lötstelle gibt eine Wärmemenge $\Pi_0 I$ ab, während in jedem Leiterelement mit einer Temperaturdifferenz dT bei der Temperatur T eine Wärmemenge $\tau I\, dT$ erzeugt wird.

Wegen der Reversibilität der Prozesse muß die gesamte Entropieänderung verschwinden:

$$\frac{\Pi_1}{T_1} I - \frac{\Pi_0}{T_0} I + I \int_{T_0}^{T_1} \frac{\tau_1 - \tau_2}{T} dT = 0.$$

Kürzen wir diese Gleichung durch I und differenzieren sie nach T[1]), so ergibt sich:

$$\frac{d\,\frac{\Pi}{T}}{dT} + \frac{\tau_1 - \tau_2}{T} = 0 .$$

$$\frac{T\,\frac{d\Pi}{dT} - \Pi}{T^2} = \frac{\tau_2 - \tau_1}{T},$$

also

$$\frac{d\Pi}{dT} - \frac{\Pi}{T} = \tau_2 - \tau_1. \tag{5}$$

In Verbindung mit der Gleichung (4) erhalten wir daraus:

$$\alpha = \frac{\Pi}{T} . \tag{6}$$

Jede der Größen in Formel (6) läßt sich einzeln messen; man kann also dadurch diese Formel nachprüfen, die sich durch Abtrennung der reversiblen von den irreversiblen Prozessen in unserem Thermoelement ergeben hat. Das heute vorliegende experimentelle Material hat die Beziehung (6) innerhalb der Fehlergrenzen des Experiments immer bestätigt[2]). Wir können also diese Beziehung, wenn wir sie als das Ergebnis eines Experiments ansehen, weiterbenutzen. Es folgt daraus

$$\frac{d\alpha}{dT} = \frac{T\,\frac{\partial \Pi}{\partial T} - \Pi}{T^2} = \frac{1}{T} \cdot \frac{\partial \Pi}{\partial T} - \frac{\Pi}{T^2} = \frac{1}{T}\left(\frac{\partial \Pi}{\partial T} - \frac{\Pi}{T}\right).$$

Der Ausdruck in Klammern ist nach (5) gleich $\tau_2 - \tau_1$. Damit erhalten wir:

$$\frac{d\alpha_{1-2}}{dT} = \frac{1}{T}(\tau_2 - \tau_1) . \tag{7}$$

In (7) steht links eine Größe, die sich auf die Grenzfläche der beiden Leiter bezieht, rechts die Differenz zweier Glieder, deren jedes von den Eigenschaften nur eines Leiters abhängt.

Man kann nun fragen, ob sich α_{1-2} nicht als Differenz einer Größe α_1, die dem ersten Leiter zuzuordnen ist, und einer Größe α_2, die nur vom zweiten Leiter abhängt, darstellen läßt. Durch Integration der Gleichung (7) erhalten wir

$$\alpha_{1-2} = \int_0^T \frac{\tau_2}{T}\,dT - \int_0^T \frac{\tau_1}{T}\,dT ,$$

[1]) Gemeint ist: Man setze $T_1 = T_0 + \Delta T$ und lasse $\Delta T \to 0$ gehen. Π_1 bzw. Π_0 sind hier Abkürzungen für $\Pi_{1-2}(T_1)$ bzw. $\Pi_{1-2}(T_0)$; analog ist in den folgenden Formeln $\Pi = \Pi_{1-2}(T)$. — Anm. d. Red.

[2]) Steenbeck und Baranski, Журнал Технической Физики (Journ. techn. Physik) *26*, 683 (1956), glauben kleine Abweichungen gefunden zu haben, die außerhalb der Meßfehler liegen. — Anm. d. Red.

wobei die untere Integrationsgrenze dem absoluten Nullpunkt entspricht, bei dem sowohl α_{1-2} als auch $\frac{d\alpha_{1-2}}{dT}$ verschwinden. Man setzt nun naheliegenderweise

$$\alpha_{1-2} = \alpha_2 - \alpha_1 = \int_0^T \frac{\tau_2}{T}\, dT - \int_0^T \frac{\tau_1}{T}\, dT,$$

$$\alpha_1 = \int_0^T \frac{\tau_1}{T}\, dT, \qquad \alpha_2 = \int_0^T \frac{\tau_2}{T}\, dT. \tag{7a}$$

Zur praktischen Bestimmung von α_1 und α_2 müßte man $\tau(T)$ vom absoluten Null punkt bis zu der gegebenen Temperatur kennen. Messungen der THOMSON-Wärme in einem so großen Temperaturgebiet sind sehr schwierig und wurden bisher nur an einer metallischen Silberlegierung durchgeführt. Die Temperaturabhängigkeit der Differenz $\alpha_2 - \alpha_1$ wurde dagegen bei sehr verschiedenartigen Materialien mehrfach gemessen.

Man sieht leicht folgendes ein: mißt man bei einer Temperatur T_0 die Differenzen $\alpha_{1-2}, \alpha_{2-3}\, \alpha_{3-4}, \ldots, \alpha_{(n-1)-n}$, so erhält man auch die Größe

$$\alpha_{1-2} + \alpha_{2-3} + \ldots + \alpha_{(n-1)-n} = \alpha_{1-n}.$$

Da sich nämlich alle Kontaktstellen 1—2, 2—3, ... auf der gleichen Temperatur T_0 befinden, hätte eine Änderung der EMK des Paares 1, n beim Zwischenschalten der Leiter 2, 3 ... zur Folge, daß Wärme vom umgebenden Medium ohne jede Kompensation in elektrische Energie überginge, und das widerspräche dem zweiten Hauptsatz der Thermodynamik. Die Unabhängigkeit der Werte α_{1-n} von Zwischengliedern ist eine Grundtatsache, die zum Aufbau der thermoelektrischen Reihe benutzt wird. Die Erfahrung hat diese Beziehung immer bestätigt.

Könnten wir einer Substanz einen bestimmten Wert $\alpha = \alpha_0$ zuschreiben, so hätten wir damit die Möglichkeit, absolute, nicht nur relative Werte für die Thermokraft jeder Substanz anzugeben. Die Gleichung (7a) gibt eine solche Möglichkeit, sobald $\int_0^T \frac{\tau}{T}\, dT$ an einer Substanz gemessen worden ist; die Genauigkeit dieser Messungen ist aber gering, obwohl sich viele Experimentatoren um die Messung von $\tau(T)$ in der Nähe des absoluten Nullpunktes bemüht haben (BORELIUS, KEYES u. a.). Einen anderen Weg zur Ermittlung von Absolutwerten von α bietet eine Betrachtung des Mechanismus der Thermoelektrizität, zu der wir im nächsten Paragraphen übergehen wollen.

Die Gleichung (6) wurde an den verschiedensten Materialien mehrfach nachgeprüft und immer bestätigt[1]). Mit Hilfe der Gleichungen (6) und (7) kann man den

[1]) S. Anm. 2 auf der vorigen Seite.

PELTIER- und den THOMSON-Koeffizienten (Π und τ) durch α und $\frac{d\alpha}{dT}$ ausdrücken. Ist speziell $\frac{d\alpha}{dT} = 0$, so ist $\tau_1 = \tau_2$, und folglich sind die THOMSON-Wärmen, die in den beiden Zweigen des Thermoelements erzeugt bzw. verbraucht werden, einander entgegengesetzt gleich. Die in einem geschlossenen Kreis auftretenden THOMSON- bzw. PELTIER-Wärmen sind

$$q = (\tau_1 - \tau_2)\, I\, (T_1 - T_2)$$

bzw.

$$Q = \alpha T I.$$

Je größer $T_1 - T_2$ gegenüber T ist, desto mehr spielt die THOMSON-Wärme eine Rolle.

Wie es scheint, kann man durch rein thermodynamische Überlegungen nicht nur den Zusammenhang zwischen Π, τ und α ermitteln, sondern auch den Wert von α selbst bestimmen. Betrachten wir nämlich einen Kontakt zwischen zwei elektronischen Halbleitern, zwischen denen — bei der konstanten Temperatur T des ganzen Kreises — ein Coulomb Elektrizität so langsam übergeht, daß I beliebig klein ist. In diesem Kreis gibt es weder Wärmeleitung noch Verluste durch JOULEsche Wärme, und der ganze Kreis kann als Gleichgewichtssystem aufgefaßt werden, in dem alle Prozesse ihr Vorzeichen wechseln, wenn man die Stromrichtung umkehrt.

Das Gleichgewicht zweier Leiter (1 und 2) setzt Gleichheit ihrer freien Energien oder chemischen Potentiale μ voraus. Wir beziehen μ auf ein Elektron und setzen es gleich der freien Energie $\bar{\varepsilon} - Ts$:

$$\mu = \bar{\varepsilon}_1 - Ts_1 = \bar{\varepsilon}_2 - Ts_2 = \bar{\varepsilon} - Ts.$$

Andererseits sind die mittleren Energien $\bar{\varepsilon}$ der Elektronen in zwei miteinander in Berührung stehenden Halbleitern verschieden, und folglich sind auch ihre Entropien s verschieden, nämlich

$$s_1 - s_2 = \frac{\bar{\varepsilon}_1 - \bar{\varepsilon}_2}{T}.$$

Geht ein Elektron durch die Grenzfläche, so ändert sich seine Energie im Mittel um $\bar{\varepsilon}_1 - \bar{\varepsilon}_2$. Diese Energie erscheint als erzeugte PELTIER-Wärme Π_{1-2}. Es ist also

$$\frac{s_1}{e} - \frac{s_2}{e} = \frac{\Pi_{1-2}}{T} = \alpha_{1-2}\,. \tag{8}$$

Man könnte also annehmen, daß die Thermokraft α an der Grenze zweier elektronischer Leiter als Differenz der Entropien einer Elektrizitätsmenge von 1 Coulomb vor und nach dem Übergang von dem einen in den anderen der beiden Leiter definiert werden kann. Diese Überlegung trifft aber nicht ganz zu: Im Gleichgewichtszustand sind die Elektronengeschwindigkeiten nach einem bestimmten statistischen Gesetz verteilt. Im Elektronenstrom stellt sich eine neue Geschwindigkeitsverteilung ein; die schnellen Elektronen etwa gehen in größerer Anzahl von einem Leiter in den anderen Leiter über als die langsamen. Zwischen den im ersten Leiter zurückgebliebenen Elektronen stellt sich dann ein neues statistisches Gleichgewicht ein, was eine weitere Entropieänderung zur Folge hat. Die von den Elektronen von dem einen Körper in den anderen transportierte Entropiemenge ist also nicht gleich der Differenz ihrer Entropien in den beiden Körpern vor dem Übergang. Die Gleichung (8) kann nur als Näherung betrachtet werden. Eine genaue Bestimmung von α_{1-2} ist nur auf Grund einer Betrachtung des Übergangsprozesses der Elektronen möglich.

Wir bezeichnen mit j den Elektronenstrom durch die Grenzfläche der beiden Körper; er setzt sich zusammen aus den Beiträgen der einzelnen Elektronen mit der Energie ε:

$$j = \int_0^\infty j(\varepsilon) d\varepsilon .$$

Die mittlere Energie der Elektronen im Strom ist

$$\bar{\varepsilon} = \frac{\int_0^\infty \varepsilon j(\varepsilon)\, d\varepsilon}{\int_0^\infty j(\varepsilon)\, d\varepsilon} .$$

Bezeichnen wir mit $f_0(\varepsilon)$ die Gleichgewichts-Energieverteilung für Elektronen, mit $l(\varepsilon)$ die freie Weglänge eines Elektrons mit der Energie ε, so läßt sich der Strom j folgendermaßen darstellen

$$j = C \int_0^\infty \frac{df_0(\varepsilon)}{d\varepsilon}\, l(\varepsilon)\, \varepsilon\, d\varepsilon .$$

Dieser Ausdruck gilt auch für einen Strom von Gasmolekülen. Setzen wir ihn[1]) in die Gleichung für $\bar{\varepsilon}$ ein, so erhalten wir

$$\bar{\varepsilon} = \frac{\int_0^\infty \frac{df_0(\varepsilon)}{d\varepsilon}\, l(\varepsilon)\, \varepsilon^2\, d\varepsilon}{\int_0^\infty \frac{df_0(\varepsilon)}{d\varepsilon}\, l(\varepsilon)\, \varepsilon\, d\varepsilon} .$$

Ist s die Entropie pro Elektron, so können wir die beim Übergang von 1 Coulomb transportierte Entropie, durch die α gegeben wird, folgendermaßen schreiben:

$$\alpha = \frac{s}{e} = \frac{1}{e}\left(\frac{\bar{\varepsilon}}{T} - \frac{\mu}{T}\right) = \frac{k}{e}\left(\frac{\bar{\varepsilon}}{kT} - \frac{\mu}{kT}\right).$$

Mit den Bezeichnungen $\varepsilon/kT = \varepsilon^*$ und $\mu/kT = \mu^*$ wird

$$\alpha = \frac{k}{e}\left[\frac{\int_0^\infty \frac{df_0(\varepsilon^*)}{d\varepsilon^*}\, l(\varepsilon^*)\, \varepsilon^{*2}\, d\varepsilon^*}{\int_0^\infty \frac{df_0(\varepsilon^*)}{d\varepsilon^*}\, l(\varepsilon^*)\, \varepsilon^*\, d\varepsilon^*} - \mu^*\right]. \qquad (8a)$$

[1]) Genauer: den $j(\varepsilon)\, d\varepsilon$ bedeutenden Integranden. — Anm. d. Red.

Die PELTIER-Wärme ist

$$\Pi = \alpha T = \frac{1}{e}(\bar{\varepsilon} - \mu).$$

Diese ganz allgemeinen Ausdrücke für Thermokraft und PELTIER-Koeffizienten gelten sowohl für entartete als auch für nichtentartete Halbleiter und Metalle.

3. Statistische Theorie.

Zu Beginn unseres Jahrhunderts ging die Metalltheorie von der Vorstellung aus, daß es in jedem Metall eine bestimmte, aber für die einzelnen Metalle verschiedene Konzentration n freier Elektronen gebe, die sich innerhalb des Metalls ähnlich wie Gasmoleküle ungeordnet bewegten. Nach diesen Theorien (RIECKE, DRUDE, LORENTZ, DEBYE) gehorchen die Elektronen ebenso der BOLTZMANN-Statistik wie die Gasmoleküle. Ihre mittlere Translationsenergie beträgt ebenso wie in einem Gas $3/2 \cdot kT$, wobei k die BOLTZMANN-Konstante, T die absolute Temperatur ist.

Die Thermospannung V in einer offenen Kette aus zwei Metallen ergibt sich nach dieser Theorie zu

$$V = \frac{k}{e}\int_{T_1}^{T_2} \ln\frac{n_2}{n_1}\, dT.$$

Wie man leicht ausrechnet, erhält man beispielsweise für $n_2/n_1 = 2{,}7$: $V = 86 \cdot 10^{-6}$ $(T_2 - T_1)$ Volt und $\alpha = 86\,\mu$V/Grad, also wesentlich mehr als in den Fällen tatsächlich beobachtet wird, in denen man ein Verhältnis $n_2/n_1 > 2{,}7$ erwarten kann.

Bei den meisten Metallen ist α nur einige μ V/Grad.

Die Quantentheorie der Metalle, die in den Jahren 1927/28 von A. SOMMERFELD einerseits und J. I. FRENKEL andererseits entwickelt wurde, hat auch diese Schwierigkeiten behoben.

Nach der Theorie von SOMMERFELD, in der auf die Elektronen im Metall die FERMI-DIRAC-Statistik angewendet wird, ist die Thermospannung zwischen zwei Metallen in erster Näherung (bei ausschließlicher Berücksichtigung der linearen Glieder) gleich Null. Erst höhere Näherungen führen zu einem endlichen, aber immer noch sehr kleinen Wert α. Das ist auch nicht verwunderlich: betrachten wir einen Metallstab, zwischen dessen Enden eine Temperaturdifferenz herrscht. Da die Temperatur die Elektronenkonzentration n im Metall nicht ändert und ihre thermischen Geschwindigkeiten nur wenig beeinflußt, so ist klar, daß in einem solchen Metall keine große EMK auftreten kann.

In der Quantentheorie ist ein Halbleiter dem klassischen Metallmodell ähnlich, wie es zu Beginn des 20. Jahrhunderts aufgestellt wurde. In Halbleitern ist nur ein Teil der Elektronen frei. Die Konzentration dieser freien Elektronen ist verschieden und gewöhnlich viel kleiner als in Metallen, so daß die klassische BOLTZMANN-Statistik anwendbar ist. Mit zunehmender Temperatur ändert sich die Konzentration der freien Elektronen und ihre kinetische Energie, die man, ebenso wie für ein Gasmolekül, gleich $3/2 \cdot kT$ setzen kann.

Wir betrachten einen Stab aus einem Halbleiter, in dem ein Temperaturgradient $\frac{dT}{dx}$ herrscht. Am heißen Ende sind Konzentration und Geschwindigkeit der Elektronen größer als am kalten. Die Elektronen diffundieren also in Richtung des Temperaturgradienten in größerer Anzahl als in umgekehrter Richtung. Der Diffusionsstrom, der negative Ladung vom heißen zum kalten Ende transportiert, erzeugt eine Potentialdifferenz zwischen diesen Enden.

Der Diffusionsprozeß wird durch das elektrische Feld, das sich im Halbleiter aufbaut, immer mehr abgebremst, bis der Elektronenstrom durch Diffusion gleich dem Rückstrom infolge der Potentialdifferenz wird. Unter diesen Bedingungen stellt sich ein stationäres Gleichgewicht der Elektronen im Halbleiter ein, bei dem die Temperaturdifferenz zwischen den beiden Enden eine entsprechende Potentialdifferenz aufrechterhält.

Durch jeden Querschnitt des Leiters tritt dann in der Zeiteinheit die gleiche Anzahl Elektronen in beiden Richtungen hindurch. Die Geschwindigkeit der Elektronen, die vom heißen Ende herkommen, ist aber größer als die Geschwindigkeit der Elektronen, die den Querschnitt in umgekehrter Richtung durchsetzen. Auf Grund dieses Unterschiedes erfolgt ein ständiger Transport von Wärmeenergie in Richtung des Temperaturgradienten, ohne daß Ladung transportiert wird.

Der Mechanismus des Wärmetransports ändert sich wesentlich, wenn sich Ladungsträger sowohl negativen (Elektronen) als auch positiven Vorzeichens (Löcher) an der Leitung beteiligen. Der gleichzeitige Transport durch beide Trägersorten in gleicher Anzahl bedeutet keine Ladungsanhäufung und deshalb auch keine Potentialdifferenz. Die gleichzeitige Diffusion von Elektronen und Löchern vom heißen zum kalten Ende beruht nicht nur auf der verschiedenen Geschwindigkeit der Ladungen, sondern auch auf dem Konzentrationsgradienten.

Bei einer solchen ambipolaren Diffusion kann ebenfalls eine Thermospannung auftreten; dies hängt von folgenden Umständen ab:

1) Wenn die Konzentration einer Ladungsträgersorte (etwa der negativen) größer ist als die der anderen, so transportiert der Strom zum kalten Ende vorzugsweise Ladung dieses Vorzeichens; hierbei wird die Bewegung der überwiegend vorhandenen Ladungsträger gebremst, die der anderen beschleunigt, bis beide Ströme gleich werden. Als Folge des Temperaturgradienten baut sich so im Leiter ein elektrisches Feld auf.

2. Eine zweite Ursache für eine Thermo-EMK sind Unterschiede in der Beweglichkeit der Ladungsträger. Die Beweglichkeit u hängt mit dem Diffusionskoeffizienten D über die EINSTEINsche Beziehung

$$\frac{u}{D} = \frac{e}{kT}.$$

zusammen.

Unter dem Einfluß des durch den Temperaturgradienten hervorgerufenen Konzentrationsgradienten würden die Ladungen mit den größten Werten von u und D am meisten beschleunigt werden, wenn sie mit zunehmender Entfernung von den entgegengesetzten Ladungen nicht eine Raumladung und damit ein elektrisches

Feld E erzeugen würden, das ihre Bewegung bremst und die zurückgebliebenen Ladungen des entgegengesetzten Vorzeichens beschleunigt. Das elektrische Feld E gleicht die Geschwindigkeiten der beiden Ladungsträgersorten aus und sorgt dafür, daß sie gemeinsam diffundieren.

Selbst dann, wenn thermische Anregung eine gleichgroße Anzahl von Ladungsträgern beider Vorzeichen erzeugt, kann also ihre Diffusion bei verschiedenen Beweglichkeiten ein elektrisches Feld in einem Leiter hervorrufen. Dieses elektrische Feld E läßt sich ähnlich ausdrücken wie bei der Ionendiffusion in einem Elektrolyten:

$$E = E_0 \frac{u_1 - u_2}{u_1 + u_2}.$$

Hierbei ist E_0 das elektrische Feld, das entstehen würde, wenn die Ladungsträger alle das gleiche Vorzeichen hätten; u_1 und u_2 sind die Beweglichkeiten der positiven und negativen Ladungen.

In einem Halbleiter nehmen sowohl die Konzentration n als auch die kinetische Energie der Ladungsträger mit der Temperatur zu. Der erste Umstand führt zum Auftreten von Potentialdifferenzen: der Konzentrationsgradient liefert

$$\alpha = \frac{kT}{e} \cdot \frac{u_1 - u_2}{u_1 + u_2} \cdot \frac{1}{n} \cdot \frac{dn}{dT},$$

wobei α verschwindet, wenn $u_1 = u_2$ oder wenn $dn/dT = 0$ ist. Der zweite Umstand — die Zunahme der Geschwindigkeit v der Ladungsträger mit der Temperatur nach der Formel $^1/_2 \cdot mv^2 = {^3/_2} \cdot kT$ — ist die Hauptursache für eine Thermo-EMK in Halbleitern mit einheitlichem Vorzeichen der Ladungsträger.

Zur Bestätigung dieser Aussage erwähnen wir die Halbleiter, bei denen eine — verglichen mit den Metallen — geringe Elektronenkonzentration innerhalb eines großen Temperaturintervalls konstant bleibt. Von dieser Art sind beispielsweise Bleisulfid und Bleitellurid mit Bleiüberschuß. Trotz der Konstanz der Konzentration liegen die Thermokräfte in der Größenordnung 100 μV/Grad, sind also ebenso groß wie bei Halbleitern mit stark temperaturabhängiger Konzentration.

Nach dem zweiten Hauptsatz der Thermodynamik können in einem geschlossenen Kreis aus elektrischen Leitern weder elektromotorische Kräfte noch Ströme existieren, falls sich der ganze Kreis auf der gleichen Temperatur befindet. An der Trennstelle zweier verschiedener Halbleiter erzeugt eine Konzentrationsdifferenz der freien Elektronen eine entsprechende Änderung des Kontaktpotentials, und dieses sorgt für das Elektronengleichgewicht im ganzen System. Diese Überlegung gilt nicht mehr, wenn nicht nur die Elektronenkonzentrationen, sondern auch die Temperaturen der Halbleiter verschieden sind, wodurch sich deren potentielle Energie ändert. Der Einfluß der Konzentration kann dabei jedoch als sekundär betrachtet werden. (Etwas andere Verhältnisse treten nur dann ein, wenn sich mit der Temperatur nicht nur die Elektronenkonzentration, sondern auch das Verhältnis zwischen den Konzentrationen negativer und positiver Ladungsträger ändert.) Ein Unterschied in den Elektronengeschwindigkeiten zwischen dem heißen und dem kalten Ende eines Halbleiters dagegen führt zu einer Potentialdifferenz zwischen diesen Enden. Je größer die

Elektronenkonzentration n im Halbleiter ist, desto kleiner ist das elektrische Feld E, das erforderlich ist, um die gleiche Anzahl von Elektronen zu transportieren, die infolge der Geschwindigkeitsdifferenz vom heißen zum kalten Ende des Stabes diffundiert.

Man kann die Thermokraft α als die beim Übergang einer elektrischen Ladung von 1 Coulomb transportierte Entropie auffassen. Wie wir sahen, hängt α nicht nur von der Entropiedifferenz in den beiden Substanzen oder den beiden Teilen eines Leiters mit verschiedenen Temperaturen, sondern auch von den Bewegungsbedingungen der Elektronen ab. Diese Bedingungen können ihrerseits von der Natur des Halbleiters, speziell dem Streumechanismus der Elektronen bei ihrer Wanderung von einem Teil des Halbleiters zum anderen abhängen. Die Größe α hängt also eng mit der Beweglichkeit u zusammen, die ebenfalls durch den Streumechanismus bestimmt wird.

In anisotropen Kristallen (mit Ausnahme kubischer Kristalle) sind der Streumechanismus und damit auch die Größen u und α von der Richtung abhängig. Man beobachtet daher zwischen zwei Stäben des gleichen Kristalls, die in verschiedenen kristallographischen Richtungen herausgeschnitten sind, eine Thermospannung, und beim Stromdurchgang wird PELTIER-Wärme erzeugt. Sind innerhalb eines homogenen Kristalls die elektrischen Stromlinien gekrümmt, so wird überall, wo eine Krümmung vorhanden ist, PELTIER-Wärme frei. Dies ist der sogenannte BRIDGMAN-Effekt.

Die für alle Anwendungen der Thermoelektrizität wichtigste experimentelle Tatsache ist wohl die Existenz von Halbleitern mit Ladungsträgern verschiedenen Vorzeichens. Wie das Experiment zeigt, entsteht bei der Erwärmung eines Halbleiterstabes an einem Ende und bei Kühlung des anderen Endes in einigen Substanzen ein elektrisches Feld, das vom heißen zum kalten Ende gerichtet ist, in anderen dagegen ein Feld von umgekehrter Richtung; es sieht so aus, als ob im ersten Fall negative Elektronen, im zweiten dagegen irgendwelche positive Ladungen vom heißen zum kalten Ende diffundierten. Dieser Effekt wird von der Quantentheorie als Folge einer unvollständigen Besetzung aller normalen Quantenzustände mit Elektronen gedeutet. Die freien Quantenzustände verhalten sich wie Elektronen, die positive Ladung haben. Für sie hat sich der Ausdruck „Löcher“ eingebürgert.

Das Wort „frei“ mit Bezug auf Elektronen und Löcher ist nicht buchstäblich aufzufassen. Die Bewegung dieser Träger im Kristallgitter ähnelt nur der Bewegung von Ladungen im Vakuum. Der Einfluß der starken elektrischen Felder der Atome, Ionen und Dipole des Kristallgitters ändert aber die Bewegungsgesetze ab. In der Bändertheorie der Halbleiter wird gezeigt, daß in zwei Fällen, nämlich

1. wenn die Elektronen im „Leitfähigkeitsband“ nur wenige Quantenzustände einnehmen und

2. wenn das normale „Valenzband“ nur wenige freie Zustände (Löcher) enthält,

die mittlere Geschwindigkeit der Elektronen oder Löcher den für die Bewegung freier Ladungen im Vakuum gültigen Gesetzen der Elektrodynamik genügt, nur mit dem Unterschied, daß für die Masse dieser Ladungen ein anderer als der übliche Massenwert des freien Elektrons ($9{,}1 \cdot 10^{-28}$ g) einzusetzen ist. (Die Ladung e der Ladungsträger im Kristall ist $-1{,}6 \cdot 10^{-19}$ Coulomb für Elektronen und $+1{,}6 \cdot 10^{-19}$ Coulomb für Löcher.) So ist beispielsweise die Beschleunigung a, die ein elektrisches Feld E hervorruft, nicht gleich eE/m, wo m die Masse des freien Elektrons ist. Man kann aber innerhalb einer ganzen Anzahl von Elementarzellen des Kristalls eine Beschleunigung $\bar{a}$ annehmen, die proportional der wirkenden Kraft ist:

$$\bar{a} = eE/m^*.$$

Die Größe m^* ist ungleich m und wird als „effektive Masse" bezeichnet. Das Verhältnis m^*/m hängt von der Struktur des Kristalls und von seiner Dielektrizitätskonstanten, bei anisotropen Kristallen auch von der Bewegungsrichtung ab.

Man darf jedoch nicht glauben, daß innerhalb des Kristallgitters das Elektron eine andere reale Masse, ein anderes Gewicht oder eine andere mit ihm verbundene Nullpunktsenergie $U = mc^2$ habe. Weder die Trägheit noch das Gewicht der Elektronen sind im Metall anders als im freien Zustand; diese Größen bestimmen die reale Masse jedes Körpers. Dadurch, daß der Halbleiter Elektronen mit sehr hoher „effektiver" Masse enthält, wird sein eigenes Gewicht nicht vergrößert.

Entsprechend bedeutet auch die Bewegung der positiven Löcher nicht, daß der Körper wirkliche freie positive Ladungen enthält. Versetzt man den Körper in schnelle Rotation, so kann man die Trägheit der darin enthaltenen freien Ladungen beobachten. Diese Ladungen erweisen sich aber als negative Elektronen mit dem gleichen Verhältnis $e/m = 1{,}76 \cdot 10^8$ Coulomb/g, wie es freie Elektronen haben; das gilt auch für Leiter mit reiner Löcherleitung. Die Begriffe des „Loches" und der „effektiven Masse" sind also als Hilfsmittel zu betrachten, welche die komplizierten statistischen Berechnungen des Verhaltens einer großen Anzahl von Elektronen im periodischen Gitterfeld vereinfachen.

Jede Störung der Periodizität auf Grund von thermischen Bewegungen der Atome, von geometrischen Defekten oder Fremdatomen, bedeutet auch eine Störung für die ohne äußeres elektrisches Feld im Mittel gradlinige und gleichförmige Bewegung des Elektrons im Kristall. Jede solche Störung bedeutet eine gewisse Wahrscheinlichkeit für eine Abweichung des Elektrons von seiner Bewegungsrichtung, die es vor dem Zusammentreffen mit der Periodizitätsstörung hatte.

Die Gitterstörungen sind also Streuzentren für die Elektronen. Die stärksten Störungen des regulären Gitters rufen eine so große Streuung hervor, daß nach der Streuung alle Bewegungsrichtungen des Elektrons mit derselben Wahrscheinlichkeit auftreten, unabhängig von seiner Bewegungsrichtung vor dem „Stoß" mit der Störung. Bei schwächeren Inhomogenitäten bestehen nur Wahrscheinlichkeiten für eine Abweichung um kleine Winkel; nach mehreren dieser Stöße ist aber wiederum die ursprüngliche Vorzugsrichtung nicht mehr spürbar. Eine Gesamtheit von Störungen, die dieser letzten Bedingung genügt, kann man als „Streuzentrum" zusammenfassen.

Der Weg eines Elektrons von einem Streuzentrum zu einem anderen oder zu einer Störung, an der das Elektron seine kinetische Energie verliert und sie an das Kristallgitter abgibt, heißt „freie Weglänge" des Elektrons bzw. des Loches. Wie die effektiven Massen sind auch die freien Weglängen für Elektron und Loch im gleichen Kristall verschieden.

Besteht der geschlossene Stromkreis eines Thermoelementes aus Halbleitern mit dem gleichen Leitungsmechanismus (entweder Elektronen- oder Löcherleitung), so sind die durch die Temperaturdifferenz erzeugten elektromotorischen Kräfte in beiden Zweigen des Thermoelements vom heißen zum kalten Ende bzw. in beiden vom kalten zum heißen Ende gerichtet. Sie wirken also im Kreis in verschiedenen Richtungen, und die Thermo-EMK des Kreises ergibt sich als ihre Differenz:

$$E = (\alpha_1 - \alpha_2)\,(T_1 - T_2).$$

Besteht dagegen der Kreis aus einem Elektronen- und einem Löcherleiter, so addieren sich die beiden Thermospannungen:

$$E = (\alpha_1 + \alpha_2)\,(T_1 - T_2).$$

Natürlich ist diese Leiterkombination vorzuziehen.

Elektronen- bzw. Löcherleiter gibt es nicht nur unter den Halbleitern, sondern auch unter den Metallen; die physikalische Bedeutung der Löcher im Metall ist aber im Bändermodell schwieriger darzustellen. Es zeigt sich, daß Metalle wie Wolfram, Molybdän, Zink u. a. eine reine Löcherleitung, Blei und Zinn eine gemischte Löcher- und Elektronenleitung haben.

Ob in einem Leiter Elektronen- bzw. Löcherleitfähigkeit[1]) vorliegt, läßt sich nicht nur aus dem Vorzeichen der Thermokraft, sondern auch aus der Richtung bestimmen, in der die Ladungsträger durch ein zur Stromrichtung senkrechtes Magnetfeld abgelenkt werden. Die Regeln des Elektromagnetismus (beispielsweise die Dreifingerregel der linken Hand, die die Richtungen von Ablenkung, Magnetfeld und Strom zueinander in Beziehung setzt), geben an, in welcher Richtung das Magnetfeld die Ladungen ablenkt. Dabei ist es gleichgültig, ob der Strom durch Elektronen von links nach rechts oder durch Löcher von rechts nach links transportiert wird. Sind die Stromträger Elektronen, so lädt sich der Teil des Leiters, in den das Feld den Strom ablenkt, gegen die gegenüberliegende Seite negativ auf; wird der Strom durch Löcher transportiert, so lädt sich die Seite, zu der hin der Strom abgelenkt wird, positiv auf.

Diese Erscheinung bezeichnet man als HALL-Effekt. Die Erfahrung lehrt, daß Messungen der Thermokraft und des HALL-Effekts stets zu übereinstimmenden Aussagen über das Vorzeichen der Ladungsträger führen, wenn der Strom nur durch Ladungen eines Vorzeichens transportiert wird.

In Leitern mit gemischter Leitfähigkeit subtrahieren sich sowohl die Thermospannungen der Löcher und Elektronen als auch die von ihnen im Magnetfeld erzeugten Potentialdifferenzen voneinander, und es ergeben sich kleine Werte. Quantitativ sind die Gesetze für diese Subtraktion bei beiden Effekten etwas verschieden. Für die Thermoelektrizität erhalten wir, wenn wir die Konzentrationen von Löchern und Elektronen mit n_+ und n_-, die Beweglichkeiten der beiden Teilchenarten mit u_+

[1]) Oft auch „Überschuß-" oder „n-Leitung" bzw. „Defekt-" oder „p-Leitung" genannt. — Anm. d. Red.

und u_- bezeichnen (unter Beweglichkeit versteht man die mittlere Driftgeschwindigkeit in einem elektrischen Feld von 1 V/cm):

$$\alpha \sim \frac{n_+ u_+ - n_- u_-}{n_+ u_+ + n_- u_-}.$$

Die elektromotorische Kraft beim HALL-Effekt ist dagegen:

$$E_H \sim \frac{u_+^2 n - u_-^2 n}{(u_+ n_+ + u_- n_-)^2}.$$

In dem Spezialfall $n_+ = n_-$ und $u_+ = u_-$ verschwinden sowohl α wie auch E_H. Man kann nun nicht mit voller Sicherheit umgekehrt schließen, bei $E_H = 0$ sei auch $\alpha = 0$; speziell bei Metallen aber kann man mit ziemlicher Genauigkeit $u_+ \cong u_-$ annehmen und aus der verschwindenden Spannung $E_H = 0$ auf $n_+ = n_-$, d. h. also auf $\alpha = 0$ schließen.

So liegen die Dinge bei den Metallen Pb und Sn, in denen die Hallspannung E_H sehr klein ist und ihr Vorzeichen beim Übergang von Pb zu Sn ändert. Dies läßt vermuten, daß auch der Fall $\alpha = 0$ zwischen α_{Pb} und α_{Sn} liegt; diese beiden Werte unterscheiden sich nur um 0,5 μ V/Grad. Die Thermokräfte, von diesem Nullpunkt an gerechnet, kann man als Absolutwerte betrachten.

Der Unterschied zwischen den Absolutwerten von α erstens nach der Definition $\alpha = \int_0^T \frac{\tau}{T} dT$ für eine Silber-Normallegierung und zweitens gemäß der Annahme, daß eine Größe zwischen α_{Pb} und α_{Sn} als Nullpunkt betrachtet werden kann, beträgt etwa 1 μ V/Grad, wie aus einem Vergleich der Tabelle von JUSTI, die die erstgenannte Definion verwendet, und der Tabelle von MEISSNER, die auf Grund der zweiten Annahme berechnet wurde, hervorgeht.

Die Thermokraft α ist auch bei Metallen sehr empfindlich gegen kleinste Verunreinigungen; die beiden thermoelektrischen Reihen differieren also nicht nur im Absolutwert von α ein wenig, sondern auch in der Reihenfolge der Metalle (vergleiche etwa die Stellung von Cu, Pt, Au Nr. 1, Nr. 2, Nr. 3; Seite 3).

Für Halbleiter spielen die Verunreinigungen eine entscheidende Rolle, wie aus der oben angegebenen zweiten MEISSNERschen Reihe folgt (s. Seite 3); hier findet man für die gleiche Verbindung sehr verschiedene Werte. So liefert die Tabelle für Cu_2O die Werte 474, 1000 und 1150 μ V/Grad, für Fe_2O_3 — 613 und —60 μ V/Grad.

Die Versetzung eines Halbleiters mit Fremdsubstanzen ist also, wie wir sehen, ein Weg zur Verbesserung seiner thermoelektrischen Eigenschaften.

4. Formeln zur Berechnung der Thermokraft

In diesem Abschnitt geben wir ohne Beweise Formeln zur Berechnung der Thermokraft α an. Ausführlichere Angaben hierzu findet man:

1. im zweiten Kapitel der Arbeit von SAMOILOWITSCH und KORENBLIT in „Успехи Физических Наук" (Band 49, Heft 2/3, 1953)[1]);

[1]) Deutsche Übersetzung in: Fortschritte der Physik *1*, 486—554 (1954) Akademie-Verlag, Berlin. — Anm. d. Red.

2. im ersten Teil der Arbeit von DAWYDOW und SCHMUSCHKEWITSCH in der gleichen Zeitschrift (Band 24, Heft 1, 1940);

3. in dem Buch von A. F. JOFFÉ „Halbleiter in der modernen Physik“ (Verlag der Akademie der Wissenschaften der UdSSR, 1954).

Die übrigen thermoelektrischen Koeffizienten — der PELTIER-Koeffizient und der THOMSON-Koeffizient — lassen sich, wie wir sahen, aus den Formeln für α bestimmen.

Für Metalle, speziell für einwertige, liefert die Quantentheorie die Beziehung:

$$\alpha = \frac{1}{3e}\pi^2 k^2 T\left[\frac{1}{\mu} + \left(\frac{1}{l}\cdot\frac{dl}{d\varepsilon}\right)\right]. \tag{9}$$

Hier ist μ das chemische Potential der Elektronen, das für Metalle mit der FERMI-Energie übereinstimmt, l ist die freie Weglänge von Elektronen mit der kinetischen Energie ε. In Metallen kann man

$$l \sim \varepsilon^2, \tag{10}$$

annehmen, und, da sich an der Leitung vorwiegend Elektronen beteiligen, für die $\varepsilon \approx \mu$ ist, erhalten wir

$$\frac{1}{\mu} + \frac{1}{l}\cdot\frac{dl}{d\varepsilon} = \frac{3}{\mu},$$

also

$$\alpha = \frac{1}{e}\pi^2 k^2 T\frac{1}{\mu} = \pi^2\frac{k}{e}\cdot\frac{kT}{\mu}. \tag{11}$$

Die Größe μ ist nahezu unabhängig von der Temperatur, so daß α in einwertigen Metallen proportional der absoluten Temperatur T wird:

$$\alpha \sim T. \tag{12}$$

Größenordnungsmäßig liefert die Formel (11)[1])

$$\alpha \cong -0.86\cdot 10^{-3}\frac{3\cdot 10^{-2}}{5} = -5.2\,\mu\text{V/Grad}$$

in guter Übereinstimmung mit den experimentellen Daten.

Eine Ausnahme macht Lithium, bei dem $\alpha = 41\ \mu$ V/Grad ist. SAMOILOWITSCH und SEROWA zeigten, daß in diesem Fall außer der Elektronendiffusion noch ein weiterer Effekt merklich wird, der darin besteht, daß die thermischen Schwingungen oder die Schallquanten (Phononen) Ladungsträger vom heißen zum kalten Ende des Leiters „mitreißen“[2]); das Glied $\frac{1}{l}\frac{dl}{d\varepsilon}$ bekommt den Wert $-2{,}8/\varepsilon$ statt $+2/\varepsilon$, der für die meisten Metalle gilt[3]). Auf die Möglichkeit einer solchen Mit-

[1]) $e < 0$. — Anm. d. Red.

[2]) In der angelsächsischen Literatur als „phonon drag“ bezeichnet. Der Effekt ist vor allem auch bei Halbleitern und tiefen Temperaturen sehr wesentlich. — Anm. d. Red.

[3]) Nach der am Anfang dieses Paragraphen zitierten Arbeit von SAMOILOWITSCH und KORENBLIT erhält man den negativen Wert $-2{,}8/\varepsilon$ für $(1/l)\,dl/d\varepsilon$ dann, wenn man verbesserte Wellenfunktionen benutzt; dann kann man wenigstens das positive Vorzeichen von α bei *Li* erklären (bei den übrigen Alkalien ist α negativ), wenn auch die quantitative Übereinstimmung mit dem Experiment noch unbefriedigend ist. Der erwähnte „Mitreiß“-Effekt hat das Auftreten eines zusätzlichen Summanden auf der rechten Seite von (9) zur Folge.

führung von Elektronen durch Schallquanten hat zuerst GUREWITSCH hingewiesen In Halbleitern muß man die Möglichkeit berücksichtigen, daß sowohl freie Elektronen mit der Konzentration n_- als auch Löcher mit der Konzentration n_+ vorhanden sind. Die von den Elektronen und den Löchern hervorgerufenen Thermokräfte haben entgegengesetztes Vorzeichen und subtrahieren sich also voneinander. In diesem Fall bekommt man für die Thermokraft:

$$\alpha = \frac{1}{T\sigma}\left[u_- n_- (2kT - \mu) - u_+ n_+ (2kT - \mu + \Delta E)\right],$$

dabei ist σ die spezifische elektrische Leitfähigkeit, u die Beweglichkeit, ΔE die energetische Breite der verbotenen Zone.

PISSARENKO leitete eine Formel für α in einer für die Rechnung sehr bequemen Form ab:

$$\alpha = \frac{k}{\sigma}\left\{ u_- n_- \left[A + \ln \frac{2(2\pi m_-^* kT)^{\frac{3}{2}}}{h^3 n_-}\right] - u_+ n_+ \left[A + \ln \frac{2(2\pi m_+^* kT)^{\frac{3}{2}}}{h^3 n_+}\right]\right\}. \tag{13}$$

Der Wert der Konstanten A hängt vom Streumechanismus der Elektronen und speziell von der Größe $\frac{1}{l}\frac{dl}{d\varepsilon}$ ab. Diese Größe wird durch den Charakter der Streuzentren bestimmt (h ist die Plancksche Konstante).

In Kristallen mit einem durch Valenzkräfte zusammengehaltenen Atomgitter besteht das Spektrum der thermischen Schwingungen aus akustischen Schwingungen, deren Frequenz zuerst (bei kleinen Frequenzen und bei den Wellenlängen, die für die Wechselwirkung mit langsamen Elektronen die Hauptrolle spielen) ihrem Impuls proportional ist. Für diese Kristalle ist

$$\frac{dl}{d\varepsilon} = 0, \quad l = \text{const.} \tag{14}$$

In Kristallen mit Ionengitter erfolgt die Elektronenstreuung vorwiegend an elektromagnetischen Polarisationsschwingungen, bei denen die positiven und die negativen Ionen sich in entgegengesetzter Richtung bewegen. Von diesen „optischen“ Schwingungen sind die Longitudinalschwingungen von der größten Bedeutung; sie erzeugen Ladungsanhäufungen und Verdünnungen. In diesem Fall entstehen sehr starke elektrische Felder, die das bewegte Elektron ablenken.

In Ionenkristallen hat man zwei Temperaturgebiete zu unterscheiden:

1. Temperaturen, bei denen die mittlere kinetische Energie der Elektronen $3/2 \cdot kT$ wesentlich kleiner ist als die Eigenschwingungsenergie der Ionen $h\nu_0$ (diese Schwingungen entsprechen den Reststrahlen); für diese Temperaturen, die unterhalb der DEBYE-Temperatur Θ [1]) liegen ($kT \ll h\nu_0$) gilt

$$l = l_0 \sqrt{\frac{\varepsilon}{h\nu_0}}\, e^{\frac{h\nu_0}{kT}} \tag{15}$$

2. bei hohen Temperaturen mit $kT \gg h\nu_0$ gilt:

$$l \sim \varepsilon. \tag{15a}$$

[1]) Definiert durch $h\nu_0 = k\Theta$. — Anm. d. Red.

In vielen Kristallen stehen bei Temperaturen oberhalb der DEBYE-Temperatur ($kT \gg h\nu_0$) die Elektronen mit Schwingungen in Wechselwirkung, deren Frequenz sich mit wachsendem Impuls der Atome wenig ändert; es ist dann

$$l \sim \varepsilon^{-1}. \tag{16}$$

Als Streuzentren für Elektronen wirken nicht nur die Schwankungen der thermischen Bewegung, d. h. die Schallquanten, sondern auch alle anderen Störungen der regulären Kristallstruktur. In erster Linie kommen hier ionisierte Fremdatome in Frage; in diesem Fall gilt wie in Metallen:

$$l \sim \varepsilon^{2}. \tag{17}$$

Die Streuung an neutralen Atomen ist im allgemeinen gering; für sie gilt:

$$l \sim \varepsilon^{\frac{1}{2}}. \tag{18}$$

Die Deformationen, die im Kristall unter dem Einfluß von Belastungen eintreten, und die geometrischen Defekte verursachen ebenfalls eine Streuung der Elektronen. Die Anzahl dieser Bereiche ist aber im allgemeinen nicht so groß, daß sie einen mit den übrigen Streuquellen vergleichbaren Einfluß ausüben.

Enthält ein Kristallgitter Streuzentren verschiedener Art, so kann man sie in guter Näherung durch einen Streuprozeß mit einer freien Weglänge ersetzen, die gemäß

$$\frac{1}{L} = \Sigma \frac{1}{l}. \tag{19}$$

zu berechnen ist[1]).

Die Konstante A in Formel (13) hängt von der Beziehung $l = l(\varepsilon)$ sowie $u = u(T)$ ab; man findet folgende Werte:

Für Atomgitter

$$l \sim \varepsilon^{0},\ u \sim \frac{1}{T^{\frac{3}{2}}},\ A = 2. \tag{20a}$$

Für Ionengitter mit $kT \ll h\nu_0$

$$l = l_0 \sqrt{\frac{\varepsilon}{h\nu_0}}\, e^{\frac{h\nu_0}{kT}}, \quad u \sim e^{\frac{h\nu_0}{kT}},\ A = 2{,}5. \tag{20b}$$

Für Ionengitter mit $kT \gg h\nu_0$

$$l \sim \varepsilon, \quad u \sim \frac{1}{T^{\frac{1}{2}}}, \quad A = 3{,}0. \tag{20c}$$

Für eine Streuung an Schwingungen konstanter Frequenz

$$\left.\begin{aligned} &l \sim \varepsilon^{-1}, \quad u \sim \frac{1}{T^{\frac{5}{2}}}, \quad A = 1{,}0 \\ &\text{oder} \\ &l \sim \varepsilon^{-\frac{3}{2}}, \quad u \sim \frac{1}{T^{3}}, \quad A = 0{,}5. \end{aligned}\right\} \tag{20d}$$

[1]) Es wird reziprok aufsummiert, da l^{-1} der Streuwahrscheinlichkeit proportional ist. — Anm. d. Red.

Für Streuung an Fremdionen

$$l \sim \varepsilon^2, \quad u \sim T^{\frac{3}{2}}, \quad A = 4{,}0. \tag{20e}$$

Allgemein erhalten wir für $l \sim \varepsilon^r$ eine Abhängigkeit $u \sim T^{r-3/2}$ und einen Wert $A = r + 2$, wenn die Streuung an thermischen Schwingungen erfolgt. Sind die Streuzentren Kristallgitterdefekte, die nicht von der Temperatur abhängen, so ist $u \sim T^{r-1/2}$.

Vom Charakter der Bindungskräfte hängt der Zusammenhang zwischen l und ε ab. Er ist schwierig zu bestimmen; man kann aber aus dem Wert r auf die Beziehung $u = u(T)$ schließen.

Bei der Anwendung der thermoelektrischen Erscheinungen in der Praxis ist es unvorteilhaft, wenn gleichzeitig Löcher und Elektronen vorhanden sind, es sei denn, die Beweglichkeiten u_+ und u_- unterscheiden sich sehr. Außerdem ist eine hohe elektrische Leitfähigkeit des Halbleiters günstig. Beide Forderungen kann man befriedigen, indem man dem Halbleiter Fremdsubstanzen zusetzt, die entweder freie Elektronen oder Löcher erzeugen.

Für einen Halbleiter mit einer Ladungsträgersorte vereinfacht sich die Formel (13) auf:

$$\alpha = \frac{k}{e}\left[A + \ln\frac{2\,(2\pi m^* kT)^{\frac{3}{2}}}{h^3 n}\right]. \tag{21}$$

Die Formeln für α können ebenfalls noch anders geschrieben werden. Für *Atomgitter* mit *einer* Ladungsträgersorte wird

$$\alpha = \frac{k}{e}\left(2 - \frac{\mu}{kT}\right). \tag{22a}$$

Wenn man mit ΔE den Abstand der Störterme von der entsprechenden Bandkante bezeichnet, gilt unter der Voraussetzung $\Delta E \gg kT$

$$\alpha \cong \frac{k}{e} \cdot \frac{\Delta E}{2kT}. \tag{22b}$$

Auf Grund der letzten Formeln müßte der Wert α mit steigender Temperatur wie $1/T$ abnehmen[1]) und bei der Annäherung an den absoluten Nullpunkt gegen Unendlich gehen. Diese Schlußfolgerung widerspricht nicht nur den experimentellen Daten, sondern auch den Sätzen der Thermodynamik, nach denen nicht nur $\alpha\,(T)_{T=0} = 0$, sondern auch $\left(\frac{d\alpha}{dT}\right)_{T=0} = 0$ sein muß.

Dieser Widerspruch weist darauf hin, daß die Näherungen bei der Herleitung der Formeln (21, 22a und 22b) unzulässig sind und daß es noch weitere unberücksichtigte Eigenschaften gibt, die bei diesen Temperaturen zum Vorschein kommen.

[1]) In (22a, b) ist die T-Abhängigkeit von μ unberücksichtigt geblieben, die auch im allgemeinen schwächer als T^{-1} ist. Den Faktor 1/2 in (22b) sollte man wohl nicht zu ernst nehmen (Näheres findet man in W. Shockley, Electrons and Holes in Semiconductors, New York 1950). Man darf sich auch nicht wundern, wenn Diskrepanzen zu experimentellen Ergebnissen auftreten; abgesehen von den eben erwähnten speziellen Voraussetzungen können auch noch andere Effekte eine Rolle spielen, wie z. B. „phonon drag" (s. Anm. S. 27). — Anm. d. Red.

Für Ionen-Halbleiter ist

$$\alpha = \frac{k}{e}\left(3 - \frac{\mu}{kT}\right). \tag{23}$$

Das PEKARsche Polaronenmodell für Ionenhalbleiter liefert eine kompliziertere Abhängigkeit $l(\varepsilon)$ und dementsprechend auch andere Ausdrücke für α. Die Theorie von PEKAR ist an Ionenhalbleitern gut bestätigt worden; nach ihr transportiert das Elektron, das ja seine Umgebung polarisiert, diesen „Polarisationszustand" ständig mit sich im Gitter herum. Für konstante Temperatur T kann man nach PEKAR setzen

$$l_{\text{pol}} \sim \varepsilon^2 .$$

In der Praxis ist es zweckmäßig, die Formel (21) als Verknüpfung der beiden leicht meßbaren Größen α und σ zu schreiben. Durch Zusatz von Fremdsubstanzen oder durch Abweichungen von der stöchiometrischen Zusammensetzung einer Verbindung kann man die Konzentration n wesentlich ändern, ohne die Beweglichkeit u zu beeinflussen. Man wird dann den Ausdruck für α folgendermaßen schreiben:

$$\alpha = \frac{k}{e}\left[A + \ln\frac{2(2\pi m^* kT)^{\frac{3}{2}} eu}{h^3}\right] - \frac{k}{e}\ln(neu) \tag{23a}$$

oder

$$\alpha = C - 86 \cdot 10^{-6} \ln\sigma = C - 2 \cdot 10^{-4} \log\sigma. \tag{24}$$

Es gibt eine größere Gruppe von Halbleitern mit hoher elektrischer Leitfähigkeit, bei denen sich die Konzentration n mit der Temperatur nicht ändert. Die effektive Masse m^* hängt ebenfalls nicht von der Temperatur ab. Für diese Materialien hat nach (21) die Temperaturabhängigkeit von α die Form

$$\alpha = \frac{k}{e}\left[A + \ln\frac{2(2\pi m^* k)^{\frac{3}{2}}}{h^3 n}\right] + \frac{3}{2}\cdot\frac{k}{e}\ln T. \tag{25}$$

In diesen Fällen werden die Formeln (24) und (25) vom Experiment gut bestätigt: In der Darstellung $\alpha = f(\log\sigma)$ bekommt man eine Gerade mit einer Steigung von nahezu $2 \cdot 10^{-4}$ (vgl. Abb. 7 und 8 S. 60 und 61). Ebenso wächst α bei konstanter Konzentration n mit zunehmender Temperatur langsam logarithmisch mit T.

Mit wachsender Konzentration freier Elektronen geht das System allmählich in den entarteten Zustand über; dann darf nicht mehr die BOLTZMANN-Statistik anstelle der allgemein für Elektronen gültigen FERMI-DIRAC-Statistik angewendet werden.

Diese allgemeineren Formeln ergeben für die Konzentration

$$n = \frac{4\pi(2m^* kT)^{\frac{3}{2}}}{h^3} F_{\frac{1}{2}}\left(\frac{\mu}{kT}\right). \tag{25a}$$

Der allgemeinste Ausdruck für α bei beliebiger Konzentration n von Ladungsträgern einheitlichen Vorzeichens hat die Form

$$\alpha = \pm\frac{k}{e}\left[\frac{r+2}{r+1}\cdot\frac{F_{r+1}\left(\frac{\mu}{kT}\right)}{F_r\left(\frac{\mu}{kT}\right)} - \frac{\mu}{kT}\right]. \tag{26}$$

Für Atomgitter, bei denen $r = 0$ ist, gilt speziell

$$\alpha = \pm \frac{k}{e} \left[2 \frac{F_1 \left(\frac{\mu}{kT} \right)}{F_0 \left(\frac{\mu}{kT} \right)} - \frac{\mu}{kT} \right]. \tag{26a}$$

In den oben angegebenen Formeln (25a, 26, und 26a) bedeuten $F_r \left(\frac{\mu}{kT} \right)$ die FERMI-Integrale

$$F_r \left(\frac{\mu}{kT} \right) = \int_0^\infty \frac{x^r dx}{e^{x - \frac{\mu}{kT}} + 1}, \tag{26b}$$

Dabei ist x die reduzierte Elektronenenergie: $x = \varepsilon / k T$.

Die Werte der Funktionen F_r für verschiedene r sowie Formeln für $\varkappa$, σ und α sind auf den Seiten 68 und 69 zusammengestellt.

SCHIFRIN diskutierte die Verhältnisse beim Übergang von der BOLTZMANN- zur FERMI-Statistik; er stellte fest, daß der Einfluß der Entartung schon bei $\mu/k T > -2$ merklich wird und daß bei $\mu/k T > 2$ das Elektronengas schon stark entartet ist.

Wie KONTOROWA zeigte, ist aber die Formel (21) so gebaut, daß sie bis zu $\mu/k T \approx 1$ keine großen Fehler liefert.

5. *Experimentelle Ergebnisse*

Eine systematische Untersuchung der Thermokräfte verschiedener Halbleiter führten HOCHBERG und SOMINSKI durch. Unter den untersuchten Stoffen war nur einer (WO_3), der der Beziehung (22b) genügte: in einem Temperaturintervall von der Temperatur der flüssigen Luft (90° K) bis 400° K nahm die Thermokraft umgekehrt proportional zur absoluten Temperatur ab.

Beim V_2O_5 wurde die durch (22b) geforderte Abhängigkeit[1]) nur oberhalb von —30° C (240° K) beobachtet; unterhalb dieser Temperatur nimmt α proportional T zu. Die Temperaturabhängigkeit von α läßt sich durch zwei sich schneidende Geraden darstellen, wenn man bis 240° K auf der Abszissenachse T, oberhalb 240° K dagegen $1/T$ aufträgt. Bemerkenswert ist jedoch, daß bei dieser Temperatur von 240° K trotz des scharfen Knicks im Verlauf von α die Temperaturabhängigkeit des Widerstandes sich nicht merklich ändert. Trägt man den $\log \varrho$ gegen $1/T$ auf, so ergibt sich im ganzen Intervall von der Temperatur der flüssigen Luft (90° K) bis 500° K eine Gerade (ϱ ist der spezifische Widerstand).

Beim Cu_2O hängt α zwischen 80 und 630° K nicht von der Temperatur ab; oberhalb 630° K nimmt es proportional zu $1/T$ ab. In diesem Fall fällt, wie ANDERSEN gezeigt hat, der Knick der Geraden α $(1/T)$ bei 630° K mit einem Knick in der Kurve $\log \varrho = f(1/T)$ zusammen. Die Konstanz von α unterhalb 630° K stimmt aber nicht mit den Voraussagen der Theorie überein, denn hier wäre eine Fremdleitung zu erwarten, für die die Formel (22b) gelten müßte.

[1]) S. Anm. S. 23.

Beim Zinnsulfid (SnS) wächst α bei niedrigen Temperaturen proportional zu T an, wird dann konstant und fällt bei hohen Temperaturen wie $1/T$ ab.

Noch komplizierter ist der Verlauf von α für Tl_2S; zunächst wächst α mit T, nimmt dann ab, wächst nochmal an und fällt bei hohen Temperaturen schließlich wieder ab.

Die hier aufgezählten Halbleiter haben verhältnismäßig hohe Widerstände und eine breite verbotene Zone. Außer ihnen liegen noch Angaben über eine ganze Reihe anderer Halbleiter vom selben Typ vor.

Die Formel (22b), nach der α mit wachsender absoluter Temperatur wie $1/T$ abnehmen sollte, läßt sich nur bei hohen Temperaturen bestätigen (bei Cu_2O oberhalb 600° K, bei BaO oberhalb 700° K, bei NiO oberhalb 700° K). Bei tiefen Temperaturen geht α im Gegensatz zu (22b), aber in voller Übereinstimmung mit den Forderungen der Quantenthermodynamik, bei Annäherung an den absoluten Nullpunkt mit T gegen Null. Dieser Verlauf von α ist für SiC unterhalb 200° K, für CdO unterhalb 700° K, für SbS unterhalb 190° K, für Ge unterhalb 30° K, ferner bei SnS, TiO_2 und einigen anderen Stoffen festgestellt worden.

Bei mittleren Temperaturen findet man häufig einen Bereich, wo α praktisch konstant bleibt; wie bereits erwähnt, trifft dies für das Cu_2O zwischen 90 und 630° K zu. Die gleiche Konstanz von α zeigt MoS_2 zwischen 300 und 500° K, Bi_2S_3 zwischen 110 und 450° K, SiC zwischen 100 und 300° K, PbS von 30 bis 750° K. In diesem Intervall halten die beiden Tendenzen — der Anstieg $\alpha \sim T$ bei tiefen Temperaturen und der Abfall $\alpha \sim 1/T$ bei hohen Temperaturen — einander sozusagen die Waage.

Die gegenwärtige[1]) Theorie der thermoelektrischen Erscheinungen führt bei Halbleitern mit einem nichtentarteten Elektronensystem nur zu einer Abnahme der Thermokraft α mit wachsender Temperatur. Eine Zunahme von α mit der Temperatur und proportional T liefert die Theorie nur für Stoffe mit entartetem Elektronengas. In den beiden Fällen (Ge und CdO), in denen diese Temperaturabhängigkeit von α bei Halbleitern festgestellt wurde, kann man zeigen, daß eine Entartung des Elektronengases wie in einem Metall vorliegt.

Es besteht jedoch kein Grund zu der Annahme, daß bei der Annäherung an den absoluten Nullpunkt die Elektronen in allen Halbleitern in den entarteten Zustand übergehen. Dagegen müssen in der Nähe des absoluten Nullpunktes α und sogar $\frac{d\alpha}{dT}$ gegen Null gehen, denn für $T \to 0°$ K verschwindet nach dem NERNSTschen Satz die Entropie S, und damit auch α.

Die experimentellen Ergebnisse stehen in guter Übereinstimmung mit diesem extrapolierten Verlauf $\alpha = f(T)$, theoretische Argumente dafür liegen aber z. Z. noch nicht vor. Es gibt auch keine allgemeine empirische Formel, welche die thermoelektrischen Eigenschaften von Halbleitern in einem großen Temperaturintervall umfaßt. Man kann nur die qualitative Regel aufstellen, nach der die Thermokraft α eines Halbleiters von dem Wert $\alpha = 0$ bei $T = 0$ zunächst proportional, dann schwächer mit der Temperatur anwächst; in einem bestimmten Temperaturintervall

[1]) Modellmäßige. — Anm. d. Red.

wird sie häufig sogar konstant und zum Schluß fällt α annähernd gemäß (22a) wieder ab.

Übrigens gehören alle Halbleiter, von denen die Rede war, zu den Stoffen mit hohem spezifischem Widerstand ($\varrho > 1\,\Omega$ cm), der für Thermoelemente mit hohem Wirkungsgrad wenig günstig ist. Angaben über Halbleiter mit geringem spezifischem Widerstand ($\varrho < 10^{-2}\,\Omega$ cm) findet man in unserer Arbeit über thermoelektrische Kühlung; ferner werden sie in diesem Buch auf den Seiten 34 bis 56 behandelt.

Zu einer genauen Berechnung von α ist die Kenntnis der effektiven Masse m^* der Ladungsträger und die des Streumechanismus erforderlich; diese Angaben sind meistens nicht genügend genau bekannt. Werte der effektiven Masse m^* kann man aus einer Reihe anderer Eigenschaften des Halbleiters erschließen, nämlich aus der Ionisierungsenergie der Störstellen, der Beweglichkeit sowie aus magnetischen und optischen Erscheinungen. Der Streumechanismus der Ladungsträger ist von großem Einfluß auf die Temperaturabhängigkeit ihrer Beweglichkeit.

Berechnet man unter Benutzung dieser Daten eines gegebenen Halbleiters dessen Thermokraft, so erhält man einen Wert α, der den Ergebnissen unmittelbarer Messungen recht nahe kommt. Eine eingehende Analyse wird in dem Buch über thermoelektrische Kühlung durchgeführt.

In den Temperaturbereichen, mit denen man es heute vorwiegend zu tun hat, ist die hier dargestellte Theorie der thermoelektrischen Erscheinungen eine geeignete Grundlage für die Rechnung. Alle so berechneten Werte von α stimmen innerhalb der Fehlergrenzen des Experiments mit der Beobachtung überein[1]).

[1]) In der Übersicht dieses Kapitels sind die bei tiefen Temperaturen an störungsarmen, genügend großen Halbleiterkristallen auftetenden Schallquanten-Mitführungseffekte (phonon-drag; s. a. Anm. 3 S. 20) nicht erwähnt; sie sind auch bei der Ableitung aller im vorstehenden angegebenen Formeln nicht mit enthalten. Diese Effekte können Werte von α bewirken, die u. U. bis zu 20mal größer sind als die nach (21) berechneten. Vgl. hierzu etwa die Arbeiten von FREDERIKSE, GEBALLE, GUREWITSCH, HERRING u. a. — Anm. d. Red.

KAPITEL II

Thermoelektrische Generatoren

6. Energetische Grundlagen der thermoelektrischen Batterien

Wir betrachten ein Thermoelement (Abb. 1), bestehend aus einem Löcher- (1) und einem Elektronenhalbleiter (2), die durch eine Metallbrücke (3) verbunden sind. An die kalten Enden ist ein Außenwiderstand R angeschlossen, der die im Thermoelement erzeugte elektrische Energie aufnimmt. Ist U die Potentialdifferenz an den kalten Enden, so beträgt die vom Thermoelement gelieferte Leistung $W = U^2/R$, und die Stromstärke ist $I = U/R$.

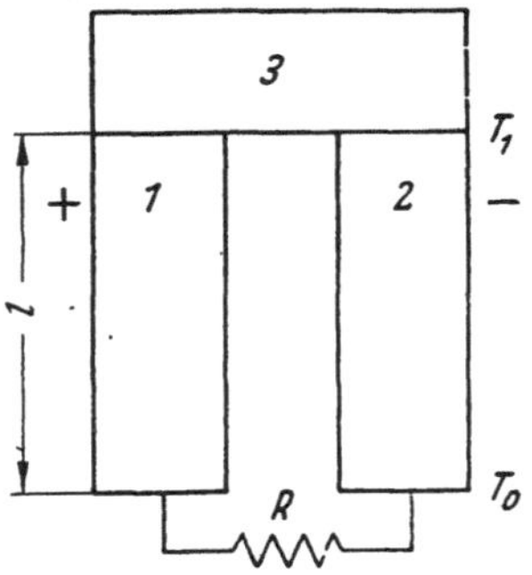

Abb. 1

Wir bezeichnen die Thermokraft, die in unserem Fall gleich der Summe der Thermokraft beider Zweige ist, mit α:

$$\alpha = \alpha_1 + \alpha_2.$$

Die durch die Brücke verbundenen heißen Lötstellen haben eine Temperatur T_1; die entsprechende thermische Energie wird von einer Quelle mit höherer Temperatur geliefert. Die kalten Enden haben eine Temperatur T_0, die ein wenig höher ist als die Temperatur der Umgebung oder einer speziellen Kühlvorrichtung.

Wir bezeichnen ferner die Innenwiderstände der beiden Zweige mit r_1 und r_2, ihre Wärmeleitwerte mit K_1 und K_2. Nennen wir den spezifischen Widerstand ϱ, die spezifische Wärmeleitfähigkeit $\varkappa$ und nehmen wir für beide Zweige die gleiche Länge l an, nennen wir ferner ihre Querschnittsflächen S_1 und S_2, so ist

$$r = r_1 + r_2 = \left(\frac{\varrho_1}{S_1} + \frac{\varrho_2}{S_2}\right) l,$$

$$K = K_1 + K_2 = (\varkappa_1 S_1 + \varkappa_2 S_2) \frac{1}{l}.$$

Auf Grund der allgemeinen Gesetze der thermoelektrischen Erscheinungen können wir die vom Thermoelement an den Stabenden abgegebene bzw. aufgenommene PELTIER-Wärme berechnen, ferner die THOMSON-Wärme, die innerhalb der Stäbe erzeugt oder verbraucht wird; ebenso berechnen wir die Wärme, die durch Wärmeleitung von den heißen zu den kalten Enden übergeht und die JOULEsche Wärme, die vom Strom in den Stäben erzeugt wird. Dann können wir die vom Thermoelement gelieferte nutzbare elektrische Energie bestimmen. Wir beziehen alle Energieformen auf eine Sekunde und drücken die Leistung in Watt aus. In den gleichen Einheiten drücken wir auch die rein thermischen Größen aus; beispielsweise die spezifische Wärmeleitfähigkeit $\varkappa$ in Watt/Grad cm statt in cal/Grad cm sec:

$$\varkappa \left[\frac{\text{cal}}{\text{Grad cm sec}}\right] = 4{,}19\, \varkappa \left[\frac{\text{Watt}}{\text{Grad cm}}\right].$$

Die pro Sekunde an den heißen Enden aufgenommene thermische Energie beträgt[1]):

$$Q_1 = \alpha_{(1)}\, I T_1.$$

Die von den kalten Enden abgegebene Leistung Q_0 ist

$$Q_0 = -\,\alpha_{(0)}\, I T_0.$$

Die in jedem der Stäbe erzeugte THOMSON-Wärme Q ist nach Gleichung (7)

$$Q = \pm \int_{T_0}^{T_1} T \cdot \frac{d\alpha}{dT} \cdot I \cdot dT.$$

Wenn α an beiden Enden den gleichen Wert hat, ist $Q = 0$.

Der Wärmestrom transportiert in beiden Stäben vom heißen zum kalten Ende sekundlich die Energie

$$Q_W = K\,(T_1 - T_0).$$

Die vom Strom in den beiden Stäben erzeugte Leistung ist

$$Q_J = I^2 r.$$

Die vom Thermoelement gelieferte Nutzleistung W ist

$$W = I^2 R;$$

die Stromstärke ist

$$I = \frac{\alpha\,(T_1 - T_0)}{R + r}.$$

Mit der Bezeichnung

$$\frac{R}{r} = m, \qquad (27)$$

erhalten wir

$$Q_1 = \alpha^2 T_1 (T_1 - T_0) \frac{1}{r\,(m+1)},$$

$$W = \alpha^2 (T_1 - T_0)^2 \frac{m}{r\,(m+1)^2}.$$

Wir vernachlässigen vorläufig die THOMSON-Wärme als klein gegen die übrigen Größen und haben deshalb schon $\alpha_{(1)} = \alpha_{(0)} = \alpha$ gesetzt. Wenn $\alpha_{(1)} \neq \alpha_{(0)}$ ist, hat eine Berücksichtigung der THOMSON-Wärme zur Folge, daß im Ausdruck für den Wirkungsgrad α durch den Mittelwert beider Enden $\bar{\alpha} = (\alpha_{(1)} + \alpha_{(0)})/2$ zu ersetzen ist.

[1]) Die Indizes beziehen sich im Gegensatz zu den vorhin verwendeten auf die Temperaturen T_1 bzw. T_0. — Anm. d. Red.

Von der JOULEschen Wärme $I^2 r$, die innerhalb des Thermoelements erzeugt wird, geht die Hälfte auf die heißen Lötstellen über und wird der Quelle zurückerstattet; es bleibt eine Leistung $^1/_2\, I^2 r$, die zur kalten Lötstelle wandert.

Den Nutzeffekt η definieren wir als Verhältnis der nutzbaren elektrischen Energie $I^2 R$, die im äußeren Kreis verbraucht werden kann, zu der von der Wärmequelle gelieferten Energie. Die letztere setzt sich zusammen aus der PELTIER-Wärme Q_1, der Wärmemenge Q_W, die durch Wärmeleitung zu den kalten Lötstellen wandert, abzüglich der elektrischen Energie $^1/_2\, I^2 r$, die der Wärmequelle zurückerstattet wird.

Es ist also

$$\eta = \frac{W}{Q_1 + Q_W - \frac{1}{2} I^2 r} =$$

$$= \frac{\alpha^2 (T_1 - T_0)^2 \frac{1}{r} \cdot \frac{m}{(m+1)^2}}{\alpha^2 T_1 (T_1 - T_0) \frac{1}{r} \cdot \frac{1}{(m+1)} + K (T_1 - T_0) - \frac{1}{2} \cdot \frac{\alpha^2 (T_1 - T_0)^2}{r (m+1)^2}} =$$

$$= \frac{T_1 - T_0}{T_1} \cdot \frac{\frac{m}{m+1}}{1 + \frac{Kr}{\alpha^2} \cdot \frac{m+1}{T_1} - \frac{T_1 - T_0}{2 T_1} \frac{1}{m+1}} .$$

Der Wirkungsgrad eines Thermoelements wird also vollständig bestimmt durch

a) die Temperaturen der heißen und der kalten Lötstelle;

b) die Größe Kr/α^2, die von den Eigenschaften der Materialien des Thermoelements abhängt; wir bezeichnen sie im folgenden mit $1/z$, so daß

$$z = \frac{\alpha^2}{Kr} ; \tag{28}$$

c) die Wahl des Verhältnisses $m = R/r$.

Um bei vorgegebenen Werten von α, $\varkappa$, ϱ und beliebigem Verhältnis $R/r = m$ einen möglichst hohen Wirkungsgrad zu erzielen, hat man die Querschnitte S_1 und S_2 des Thermoelementes in der günstigsten Weise zu wählen, d. h. so, daß bei den gegebenen Werten von $\varkappa$ und ϱ das Produkt Kr möglichst klein wird. Um das Minimum von Kr zu bestimmen, differenzieren wir die Größe

$$Kr = (\varkappa_1 S_1 + \varkappa_2 S_2) \left(\frac{\varrho_1}{S_1} + \frac{\varrho_2}{S_2} \right) = \varkappa_1 \varrho_1 + \varkappa_2 \varrho_2 + \varkappa_1 \varrho_2 \frac{S_1}{S_2} + \varkappa_2 \varrho_1 \frac{S_2}{S_1}$$

nach S_1/S_2 und setzen den erhaltenen Ausdruck gleich Null. Wir erhalten

$$\frac{\varrho_1}{\varkappa_1} \cdot \frac{\varkappa_2}{\varrho_2} = \left(\frac{S_1}{S_2} \right)^2 . \tag{29}$$

Bei dieser Wahl von S_1/S_2 wird

$$Kr = \left(\sqrt{\varkappa_1 \varrho_1} + \sqrt{\varkappa_2 \varrho_2} \right)^2 ,$$

$$z = \frac{\alpha^2}{Kr} = \frac{\alpha^2}{(\sqrt{\varkappa_1 \varrho_1} + \sqrt{\varkappa_2 \varrho_2})^2} \text{ Grad}^{-1} . \tag{30}$$

Dieser Ausdruck enthält nur noch die Materialeigenschaften der beiden Zweige des Thermoelements, nicht mehr aber ihre Abmessungen.

Nun bestimmen wir das vom Standpunkt des Nutzeffekts günstigste Verhältnis $R/r = m$. Fragen wir nach der *maximalen Leistung*, so führt diese Forderung bei einem Thermoelement ebenso wie bei anderen Stromquellen zur Bedingung $R = r$, also $m = 1$, und es wird

$$\eta = \frac{1}{2} \frac{T_1 - T_0}{T_1 + \frac{2}{z} - \frac{1}{4}(T_1 - T_0)};$$

den *maximalen Wirkungsgrad* erhalten wir dagegen, wenn wir $\frac{\partial \eta}{\partial m} = 0$ setzen; nach elementaren Rechnungen erhält man

$$\left(\frac{R}{r}\right)_{\text{opt}} = M = \sqrt{1 + \frac{1}{2} z(T_1 + T_0)}, \tag{31}$$

wobei $z(T_1 + T_0)$ und M dimensionslose Zahlen sind.

Setzen wir diesen Optimalwert von m — wir nennen ihn M — in den Ausdruck für η ein, so ergibt sich

$$\eta = \frac{T_1 - T_0}{T_1} \frac{M - 1}{M + \frac{T_0}{T_1}}. \tag{32}$$

Der erste Faktor stellt hier den thermodynamischen Wirkungsgrad einer reversiblen Maschine dar, der zweite gibt an, wie stark sich dieser Wirkungsgrad infolge der irreversiblen Verluste durch Wärmeleitung ($\varkappa$) und JOULEsche Wärme (ϱ), die in dem Ausdruck für z vorkommen, verringert.

Je größer M gegen Eins ist, je größer also z und $T_1 + T_0$ sind, desto geringer ist diese Herabsetzung des Wirkungsgrades. Durch eine Erhöhung der Temperatur T_1 der heißen Lötstelle wird also η nicht nur infolge der Zunahme des Wirkungsgrades $(T_1 - T_0)/T_1$ der reversiblen Maschine heraufgesetzt, sondern ebenso auch durch das gleichzeitige Anwachsen von M bei vorgegebenem Wert von z.

Diese Formel zeigt auch ganz deutlich, daß zur Erreichung des maximalen Wirkungsgrades an das Material des Thermoelements nur eine Bedingung zu stellen ist: Es muß der maximal mögliche Wert von z, der mit der höchsten Temperatur T_1 der Wärmequelle zu vereinbaren ist, erreicht werden, oder genauer gesagt: der höchste Wert des Produktes $z\,(T_1 + T_0)/2$, der für das gegebene Material möglich ist.

Hätten wir in dem Ausdruck für η die Größe $^1/_2\, I^2 r$ gegenüber $Q_1 + Q_W$ vernachlässigt, so hätten wir erhalten:

$$M' = \sqrt{1 + T_1 z}, \tag{33}$$

$$\eta' = \frac{T_1 - T_0}{T_1} \cdot \frac{M' - 1}{M' + 1}. \tag{34}$$

Der Wert η' weicht nur wenig von η ab, solange η selbst klein ist. Beispielsweise ist für $T_1 = 600°$ K, $T_0 = 300°$ K und $z = 2 \cdot 10^{-3}$: $M = 1{,}38$, $M' = 1{,}48$; $\eta = 0{,}101$, $\eta' = 0{,}097$.

Die Abhängigkeit der Größen η und M von dem Produkt $^1/_2\, z\,(T_1 + T_0)$ ist in den Abbildungen 2 und 3 und in Tabelle 1 für $T_1 = 600°$ K, $T_0 = 300°$ K, also $(T_1 - T_0)/T_1 = 0{,}5$ dargestellt.

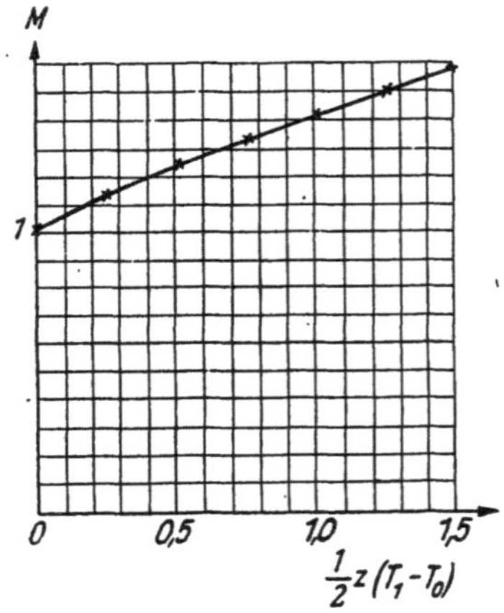

Abb. 2

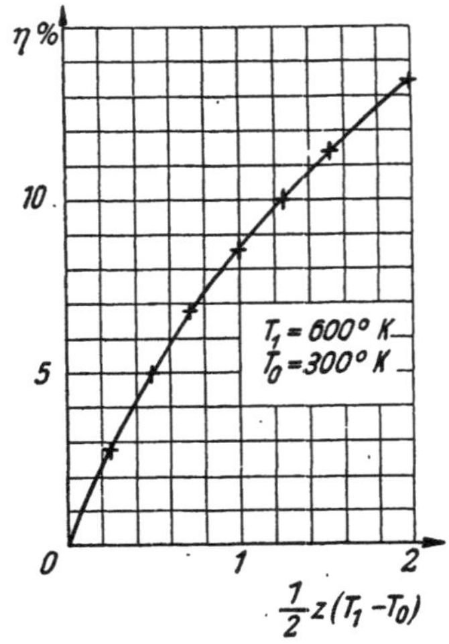

Abb. 3

Tabelle 1

$\frac{1}{2} z (T_1 + T_0)$	M	$\frac{M-1}{M + \frac{T_1}{T_0}}$	η%
0,25	1,12	0,074	3,7
0,50	1,23	0,133	6,6
0,75	1,32	0,176	8,8
1,0	1,41	0,215	10,8
1,25	1,50	0,250	12,5
1,50	1,58	0,280	14,0
2,0	1,73	0,330	16,5

Tabelle 2

T_1 °K / $z \cdot 10^3$	400		500		600		700		800		1000	
	M	η%	M	η%	M	η%	M	η%	M	η%	M	η%
0,5	1,085	1,15	1,095	2,25	1,111	3,45	1,12	4,4	1,13	5,4	1,15	7,2
0,75	1,13	1,5	1,14	3,2	1,155	4,7	1,17	6,1	1,19	7,6	1,22	10,0
1,0	1,162	2,13	1,185	4,15	1,20	5,9	1,225	7,8	1,25	9,6	1,285	12,5
1,25	1,20	2,65	1,225	5,0	1,25	7,2	1,275	9,2	1,30	11,1	1,35	15,0
1,5	1,24	3,0	1,27	5,8	1,295	8,2	1,325	10,5	1,35	12,7	1,40	16,5
1,75	1,27	3,35	1,30	6,3	1,325	9,2	1,37	11,8	1,40	14,0	1,46	18,0
2,0	1,31	3,75	1,345	7,1	1,375	10,1	1,42	13,0	1,45	15,5	1,52	20,0
2,5	1,38	4,4	1,41	8,2	1,46	11,7	1,50	14,8	1,55	17,5	1,62	22,6
3,0	1,43	4,9	1,48	9,2	1,53	13,0	1,85	16,5	1,63	19,5	1,72	25,0
4,0	1,55	6,0	1,62	11,1	1,675	15,5	1,73	19,2	1,88	23,0	1,90	28,5
5,0	1,66	6,8	1,73	12,5	1,80	17,2	1,87	21,5	1,94	25,0	2,08	32,0

Tabelle 2 und Abbildung 4 stellen für $T_0 = 300°$ K den Wirkungsgrad η in % in Abhängigkeit von z und T_1 dar.

Zur schnellen, allerdings auch weniger genauen Abschätzung des Wirkungsgrades eines Thermoelements kann man Potenzfunktionen von T_1 und z benutzen; für $T_0 = 300°$ K gilt:

$$\eta = 4 \cdot 10^{-2} T_1^{\frac{3}{2}} z^{\frac{2}{3}} .$$

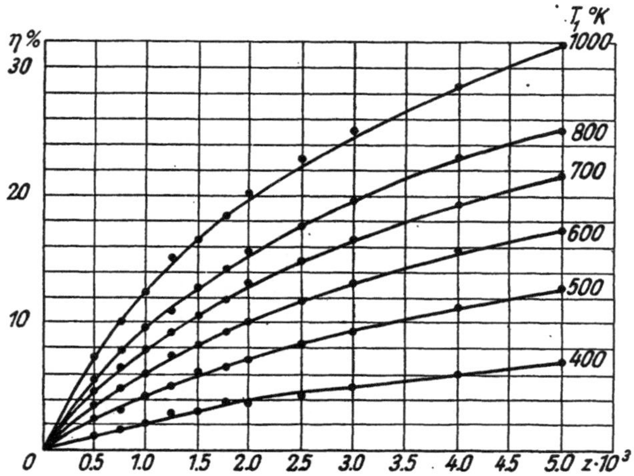

Abb. 4

Diese Werte sind in der folgenden Tabelle mit den exakten Werten

$$\eta_0 = \frac{T_1 - T_0}{T_1} \frac{\sqrt{1 + \frac{z}{2}(T_1 - T_0)} - 1}{\sqrt{1 + \frac{z}{2}(T_1 + T_0)} + \frac{T_0}{T_1}}$$

innerhalb des Intervalls $10^{-3} \leq z \leq 4 \cdot 10^{-3}$ und $500 \leq T_1 \leq 1000°$ K verglichen (letzte Spalte in %):

$z \cdot 10^3$	$z^{\frac{2}{3}} \cdot 10^2$	T_1	$T_1^{\frac{3}{2}} \cdot 10^{-3}$	$\eta = 4 \cdot 10^{-2} \cdot T_1^{\frac{3}{2}} z^{\frac{2}{3}}$	η_0	$\frac{\eta - \eta_0}{\eta_0}$
1	1	500	11,2	4,4	4,1	+7
		600	14,7	5,9	5,9	0,0
		800	22,5	9,0	9,3	−2,7
		1000	31,5	12,6	12,5	+0,8
2	1,58	500	17,5	7,0	7,0	0,0
		600	23,2	9,4	10,0	−6
		800	35,5	14,3	15,2	−6
		1000	49,5	19,7	19,6	+0,5
3	2,18	500	24,3	9,6	9,3	+3,2
		600	32,0	12,8	13,0	−1,5
		800	49,0	19,6	19,5	+0,5
		1000	68,0	27,5	25,0	+10
4	2,53	500	28,2	11,3	11,3	0,0
		600	37,0	15,0	15,0	−3,2
		800	57,0	23,0	22,5	+2,2
		1000	79,0	31,5	28,5	+ 10

Innerhalb der engeren Grenzen $500 \leq T_1 \leq 800°$ K und $1{,}5 \cdot 10^{-3} \leq z \leq 4 \cdot 10^{-3}$ kann man mit einem Fehler, der 15% nicht übersteigt, folgende Formel verwenden:

$$\eta = 6 \cdot 10^{-4}\, T^2 \sqrt{z}.$$

7. Materialien für Halbleiter-Thermoelemente

Wir wenden uns nun der Auswahl der günstigsten Materialien für Thermoelemente zu.

Vor allem müssen wir feststellen, daß Halbleiter wegen ihres verhältnismäßig großen Wertes α entschiedene Vorzüge vor Metallen haben. Bei den meisten Metallen ist nämlich $\alpha < 10 \cdot 10^{-6}$ V/Grad. Das Verhältnis der elektrischen Leitfähigkeit $\sigma = 1/\varrho$ zur Wärmeleitfähigkeit $\varkappa$ hat bei allen Metallen nahezu den theoretischen, aus der Quantenmechanik folgenden Wert[1])

$$\frac{\varkappa}{\sigma} = \frac{\pi^2}{3}\left(\frac{k}{e}\right)^2 T = 2{,}44 \cdot 10^{-8}\, T.$$

Für eine mittlere Temperatur von $T = 500°$ K wird

$$z \cong 8{,}2 \cdot 10^{-6}$$

und für $T_1 = 700°$ K, $T_0 = 300°$ K erhalten wir

$$\eta = 8{,}5 \cdot 10^{-4} = 0{,}085\,\% \,.$$

Für das günstigste Paar von Metallen, Wismut-Antimon, kann man setzen

$$\alpha = 50 \cdot 10^{-6},\quad \varkappa\varrho = 9 \cdot 10^{-6}.\quad z \cong 2{,}8 \cdot 10^{-4},\quad T_1 < 500°\text{ K},\quad T_0 = 300°\text{ K},\quad \eta = 1{,}4\%.$$

Bei einer Reihe von Halbleitern kann z die Größenordnung 10^{-3} erreichen oder sogar übertreffen. η wird dann nahezu gleich 10%. Außerdem gibt es Halbleiter mit hohen Schmelztemperaturen, die hohe Werte von T_1 zulassen. Es ist also nicht unmöglich, mit Hilfe von Halbleitern Wirkungsgrade von 10 — 15% oder sogar mehr zu erreichen. In der Literatur werden η-Werte von 3,3 bis zu 7% angegeben.

Je größer der Wert $z = \alpha^2/\varkappa\varrho$ der einzelnen Zweige des Thermoelements ist, desto größer wird auch der Wert

$$z = \frac{(\alpha_1 + \alpha_2)^2}{(\sqrt{\varkappa_1\varrho_1} + \sqrt{\varkappa_2\varrho_2})^2}$$

der den Wirkungsgrad des ganzen Thermoelements angibt. Der Zusammenhang zwischen dem Wert z des ganzen Thermoelementes und denen der beiden Zweige (z_1 und z_2) läßt sich aber nicht in allgemeiner Form angeben. In Einzelfällen gilt für $z_1 = z_2$ einfach $z = z_1 = z_2$; für $\varkappa_1\varrho_1 = \varkappa_2\varrho_2$ ist $z = 1/4 \cdot (z_1 + z_2) + 1/2 \cdot \sqrt{z_1 \cdot z_2}$. Hat man wie für die besten Halbleitermaterialien (s. unten) $\alpha_1 = \alpha_2 = 172\ \mu$V/Grad, so ist

$$z = \frac{4}{\left(\frac{1}{\sqrt{z_1}} + \frac{1}{\sqrt{z_2}}\right)^2} = \frac{4 z_1 z_2}{(\sqrt{z_1} + \sqrt{z_2})^2}\,.$$

[1]) Das sog. WIEDEMANN-FRANZsche Gesetz. — Anm. d. Red.

Mit Hilfe von (21) können wir bei Halbleitern den Zusammenhang zwischen α und $\sigma = 1/\varrho$ ermitteln. Bei konstanter Temperatur und mit der Annahme, daß sich die Beweglichkeit bei einer Änderung der Konzentration n praktisch nicht ändert, kann man einen linearen Zusammenhang zwischen α und $\ln \sigma$ erwarten[1]):

$$\alpha = C - 86 \cdot 10^{-6} \ln \sigma = C - 2 \cdot 10^{-4} \log \sigma. \tag{35}$$

Der Wert von C hängt von der Temperatur T, der Beweglichkeit und dem Charakter der chemischen Bindung des Halbleiters ab.

Andererseits hängt σ mit $\varkappa$ zusammen. Die Wärmeleitfähigkeit $\varkappa$ eines Halbleiters setzt sich zusammen aus der Wärmeleitfähigkeit der Elektronen $\varkappa_{el}$ und der Wärmeleitfähigkeit der thermischen Schwingungen und Wellen (wir bezeichnen sie als Wärmeleitfähigkeit der Schallquanten $\varkappa_s$):

$$\varkappa = \varkappa_{el} + \varkappa_s. \tag{36}$$

Die erste dieser Größen hängt mit der elektrischen Leitfähigkeit σ nach dem WIEDEMANN-FRANZschen Gesetz zusammen. Der Proportionalitätsfaktor ist aber nur bei sehr hohen Elektronenkonzentrationen ($> 2{,}5 \cdot 10^{19}$ cm^{-3}), wenn Entartung vorliegt, identisch mit dem für Metalle gültigen Wert

$$\frac{\varkappa_{el}}{\sigma} = \frac{\pi^2}{3}\left(\frac{k}{e}\right)^2 T. \tag{37}$$

Bei den geringeren Konzentrationen, mit denen man es in Halbleitern gewöhnlich zu tun hat, gilt

$$\frac{\varkappa_{el}}{\sigma} = 2\left(\frac{k}{e}\right)^2 T = 1.48 \cdot 10^{-8} \cdot T \ \mathrm{V^2/Grad}. \tag{38}$$

Für Zimmertemperatur $T = 293°$ K erhält man also

$$\frac{\varkappa_{el}}{\sigma} = 4{,}35 \cdot 10^{-6} \ \mathrm{V^2/Grad}^{2)}.$$

Mit anderen Worten: es entspricht eine Leitfähigkeit von 1000 Ω^{-1} cm^{-1} einer Wärmeleitfähigkeit von $4{,}35 \cdot 10^{-3}$ W/Grad cm oder $1{,}04 \cdot 10^{-3}$ cal/Grad cm sec.

Die Abbildung 5, die das Ergebnis unserer Messungen darstellt, veranschaulicht die Formeln (36), (37) und (38) am Beispiel eines Halbleiters, dessen Leitfähigkeit durch Änderung der Elektronenkonzentration verändert wurde.

Für $\sigma < 2500\ \Omega^{-1}$ cm^{-1} sind die Elektronen noch nicht entartet. Dann ist $\varkappa_{el} = 4{,}5 \cdot 10^{-6} \sigma$. Oberhalb $\sigma = 4000\ \Omega^{-1}$ cm^{-1} sind die Elektronen schon ziemlich stark entartet, und in diesem Gebiet nähert sich $\varkappa_{el}$ allmählich dem Wert $\varkappa_{el} = 7{,}3 \cdot 10^{-6} \sigma$.

Das WIEDEMANN-FRANZsche Gesetz gibt den Zusammenhang zwischen der Wärmeleitfähigkeit $\varkappa_{el}$ und der elektrischen Leitfähigkeit σ nur für Halbleiter

[1]) S. (24) und vorhergehende Formel. — Anm. d. Red.

[2]) Im allgemeinen Fall hat das WIEDEMANN-FRANZsche Gesetz für Halbleiter folgende Form:

$$\frac{\varkappa_{el}}{\sigma} = (r + 2)\left(\frac{k}{e}\right)^2 T. \tag{38a}$$

Dies führt für Atomgitter mit $r = 0$ zu dem Ausdruck (38). Die Größe r bedeutet ebenso wie in den Formeln (20) den Exponenten in der Potenz-Abhängigkeit der freien Weglänge l der Elektronen von ihrer kinetischen Energie.

einheitlichen Ladungsvorzeichens der Stromträger gut wieder. Sind gleichzeitig beide Sorten von Ladungsträgern vorhanden, so ist ein ständiger Ladungsstrom in Richtung des Temperaturgradienten möglich, und die Wärmeleitfähigkeit $\varkappa_{el}$ nimmt stark zu.

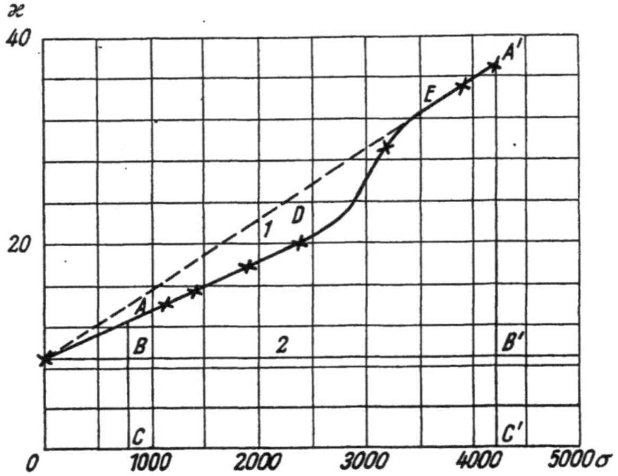

Abb. 5. Abhängigkeit der Wärmeleitfähigkeit eines Halbleiters von seiner Leitfähigkeit. 1 : Wärmeleitfähigkeit $\varkappa = \varkappa_{el} + \varkappa_s$; 2 : Wärmeleitfähigkeit $\varkappa_s$; AB und $A'B'$: Wärmeleitfähigkeit $\varkappa_{el}$; D : Beginn der Entartung; E : vollständige Entartung.

Den Schallquantenanteil der Wärmeleitfähigkeit $\varkappa_s$ kann man in Analogie zu Diffusion und Wärmeleitfähigkeit der Gase folgendermaßen schreiben:

$$\varkappa_s = \frac{1}{3} c v \bar{\lambda}. \qquad (39)$$

Dabei ist c die spezifische Wärme (in Joule/Grad cm³), v die Schallgeschwindigkeit (in cm/sec) und $\bar{\lambda}$ die mittlere freie Weglänge der Schallquanten (in cm).

Bei den meisten Halbleitern liegen diese Werte innerhalb folgender Grenzen:

$$1{,}2 \leqslant c \leqslant 1{,}6;\ 2 \cdot 10^5 \leqslant v \leqslant 5 \cdot 10^5 \text{ und } 10^5 \leqslant {}^1/_3 cv \leqslant 3 \cdot 10^5.$$

Die Wärmeleitfähigkeit $\varkappa_s$ eines Halbleiters hängt auch mit der Beweglichkeit u der Elektronen und folglich auch mit der elektrischen Leitfähigkeit zusammen. Die Beweglichkeit der Elektronen läßt sich nämlich durch die mittlere freie Weglänge $\bar{l}$ der Elektronen ausdrücken.

Für Atomgitter ist

$$u = \frac{4}{3} \frac{e}{\sqrt{2 \pi m^* kT}} \bar{l}.$$

Für Ionengitter erhalten wir bei hohen Temperaturen ($T \gg \Theta$, wobei Θ die DEBYE-Temperatur ist):

$$u = \frac{8}{3} \frac{e}{\sqrt{2 \pi m^* kT}} \bar{l}.$$

und bei tiefen Temperaturen ($T \ll \Theta$):

$$a = \frac{e \cdot e^{\frac{h\nu_0}{kT}}}{\sqrt{2 m^* h\nu_0}} \bar{l}.$$

Die elektrische Leitfähigkeit schreibt sich folgendermaßen mit u und n:

$$\sigma = eun.$$

Sowohl Schallquanten als auch Elektronen werden an den gleichen Inhomogenitäten des Kristallgitters gestreut, wenn auch in verschiedenem Maße, weil ihre

Wellenlängen ($7 \cdot 10^{-7}$ cm für Elektronen, $\sim 5 \cdot 10^{-8}$ cm für Schallquanten) und ihre physikalische Natur verschieden sind. Es muß also zwischen den Größen $\bar{\lambda}$ und $\bar{l}$ eine gewisse Korrelation bestehen. Leider ist diese Korrelation noch nicht näher untersucht worden, so daß der Zusammenhang zwischen $\varkappa_s$ und u bzw. zwischen $\varkappa_s$ und σ noch nicht formuliert werden kann[1]).

Im Falle $\varkappa_s \gg \varkappa_{el}$ ist $\varkappa$ praktisch unabhängig von der Ladungsträgerkonzentration und damit von α und σ. Geringe Störstellenkonzentrationen, die n und damit auch σ um einen Faktor von der Größenordnung 10 oder 100 heraufsetzen, haben in diesen Fällen nur einen geringen Einfluß auf u und $\varkappa$. Für solche Halbleiter hat man also nur nach dem Maximum von $\alpha^2\sigma$ zu fragen, wobei α und σ im wesentlichen als Funktionen der Konzentration n aufzufassen sind.

Wir betrachten die Bedingungen, unter denen $\alpha^2\sigma$ sein Maximum annimmt. Nach (21) ist

$$\alpha = \frac{k}{e}\left[A + \ln\frac{2(2\pi m^* k)^{\frac{3}{2}}}{h^3} + \ln e u T^{\frac{3}{2}} - \ln\sigma\right] = \frac{k}{e}(S - \ln\sigma),$$

$$\alpha^2\sigma = \left(\frac{k}{e}\right)^2 \sigma[S^2 - 2S\ln\sigma + (\ln\sigma)^2],$$

$$\frac{\partial(\alpha^2\sigma)}{\partial\sigma} = \left(\frac{k}{e}\right)^2 [(\ln\sigma)^2 - 2(S-1)\ln\sigma + S(S-2)].$$

Die Bedingung $\frac{\partial(\alpha^2\sigma)}{\partial\sigma} = 0$ liefert

$$\ln\sigma = S - 1 \pm 1.$$

Das obere Vorzeichen (+) liefert $\ln\sigma = S$ und $\alpha = 0$, d. h. $\alpha^2\sigma = 0$. Beim unteren Vorzeichen (—) wird $\ln\sigma = S - 2$ und

$$\alpha = 2\frac{k}{e} = 172 \cdot 10^{-6}\ \mathrm{V/Grad} \tag{40}$$

$$(\alpha^2\sigma)_{\max} = 3 \cdot 10^{-8} \cdot \sigma. \tag{41}$$

Die Bedingung (40) für den Maximalwert von $\alpha^2\sigma$ gilt für alle Temperaturen und für alle Werte des Gliedes A in der Formel (21). Die Größe $(\alpha^2\sigma)_{\max}$ wird also unter diesen Verhältnissen durch σ bestimmt. Die Temperaturabhängigkeit von $\alpha^2\sigma$ hat die Form

$$\alpha^2\sigma = \left(\frac{k}{e}\right)^2 \sigma(T)\left[A + \ln\frac{2(2\pi m^* k)^{\frac{3}{2}}}{h^3} + \ln\frac{e u T^{\frac{3}{2}}}{\sigma(T)}\right]^2.$$

Bei Atomgittern ist $u = u_0\, T^{-3/2}$. In diesem Fall ist

$$\alpha^2\sigma = \left(\frac{k}{e}\right)^2 \sigma(T)\left[A + \ln\frac{2(2\pi m^* k)^{\frac{3}{2}}}{h^3} e u_0 - \ln\sigma(T)\right]^2.$$

[1]) In Zusammenhang mit den früher erwähnten Effekten (s. Anm. auf S. 27) sind in jüngster Zeit theoretische Untersuchungen über die „Schallquanten"-Weglänge $\bar{\lambda}$ erfolgt (Phonon-Phonon-Streuung; Streuung an Gitterstörungen etc. und -Begrenzungen); da die „Elektronen"-Weglänge $\bar{l}$ aus der Leitfähigkeitstheorie einigermaßen bekannt ist (Elektron-Phonon-Streuung, Streuung an Gitterstörungen usw.) ist somit der vom Autor geforderte Zusammenhang wenigstens implizit bis zu einem gewissen Grade hergestellt. Einen einfachen Zusammenhang zwischen $\bar{l}$ und $\bar{\lambda}$ wird man im allgemeinen kaum erwarten dürfen. — Anm. d. Red.

Der Maximalwert von $\alpha^2\sigma$ tritt bei einem σ ein, bei dem $\alpha = 2k/e$ ist, d. h. bei

$$\ln \sigma = (A - 2) + \ln \frac{2\,(2\pi m^* k)^{\frac{3}{2}}}{h^3} + \ln e u T^{\frac{3}{2}}.$$

Für Atomgitter ist $A = 2$ und $uT^{3/2} = u_0 T_0^{3/2}$, also

$$\sigma_{\text{opt}} = \frac{2e\,(2\pi m^* k T_0)^{\frac{3}{2}}}{h^3}\, u_0 .$$

Hierbei ist u_0 die Beweglichkeit bei der Temperatur T_0. Wir setzen $T_0 = 300°\,\text{K}$ und $m^* = m_0$; dann wird

$$\frac{2\,(2\pi m_0 k T_0)^{\frac{3}{2}}}{h^3} = 2{,}5 \cdot 10^{19},$$

$$\sigma_{\text{opt}} = 2{,}5 \cdot 10^{19} \cdot 1{,}6 \cdot 10^{-19} \cdot u_0 \cong 4 u_0 \tag{42}$$

für alle Temperaturen.

Bei vielen Halbleitern, die sich in thermoelektrischen Batterien verwenden lassen, ist $n = \text{const}$ und $u \sim T^{-5/2}$ oder $uT^{5/2} = u_0 T_0^{5/2}$ und $A = 1$. Für solche Stoffe ist

$$\sigma_{\text{opt}} = \frac{1}{2{,}7} \cdot \frac{2e\,(2\pi m^* k T_0)^{\frac{3}{2}}}{h^3}\, u_0 \frac{T_0}{T}.$$

Für $m^* = m_0$ gilt

$$\sigma_{\text{opt}} = 1{,}44 \cdot u_0 \frac{T_0}{T}\, \Omega^{-1}\,\text{cm}^{-1}. \tag{43}$$

Die Optimalwerte von z in diesen Fällen sind:
für Atomgitter mit $A = 2$

$$z_{\text{opt}} = \frac{\alpha^2 \sigma}{\varkappa} = 1{,}16 \cdot 10^{-7} \cdot \frac{u_0}{\varkappa}, \tag{44}$$

für Gitter mit $A = 1$

$$z_{\text{opt}} = 4{,}3 \cdot 10^{-8} \cdot \frac{u_0}{\varkappa} \cdot \frac{T_0}{T}\,\text{Grad}^{-1}. \tag{45}$$

Für $\varkappa \cong \varkappa_s$ ist $\varkappa T \cong \text{const} = \varkappa_0 T_0$ und

$$z_{\text{opt}} = 4{,}3 \cdot 10^{-8} \cdot \frac{u_0}{\varkappa_0}\,\text{Grad}^{-1}.$$

Bei $u_0 = 1000\,\text{cm}^2/\text{V sec}$ und $\varkappa_0 = 5 \cdot 10^{-3}\,\text{cal/Grad cm sec} = 2{,}1 \cdot 10^{-2}\,\text{W/Grad cm}$ ist

$$z_{\text{opt}} = 2{,}05 \cdot 10^{-3}\,\text{Grad}^{-1}.$$

Der Minimalwert $\alpha = 0$, $\alpha^2\sigma = 0$ und $z = 0$ ergibt sich, wenn für ein Atomgitter

$$\sigma_{\min} = 29 u_0 \cdot \Omega^{-1} \cdot \text{cm}^{-1},$$

für ein Gitter mit $A = 1$

$$\sigma_{\min} = 10{,}7\, u_0 (T_0/T)$$

ist. In Abbildung 6 ist der Zusammenhang zwischen $\frac{\alpha^2\sigma}{(\alpha^2\sigma)_{\max}}$ und $\ln \frac{\sigma}{u_0}$ dargestellt.

Aus den Formeln (42) und (43) ergibt sich, daß die optimale Konzentration n der freien Elektronen bei 300° K für ein Atomgitter mit $A = 2$ folgenden Wert hat:

$$n_{\text{opt}} = \frac{\sigma_{\text{opt}}}{eu} = \frac{4u}{eu} = 2{,}5 \cdot 10^{19} \cdot \text{cm}^{-3}. \tag{46}$$

Diese Bedingung ist identisch mit der Forderung

$$n = \frac{2(2\pi m_0 kT)^{3/2}}{h^3}$$

bei der $\alpha = k/e\,(2 + 0) = 2\,k/e$ ist.

Für ein Gitter mit $A = 1$ wird bei $T = T_0 = 300°$

$$n_{\text{opt}} = 9 \cdot 10^{18} \cdot \text{cm}^{-3}. \tag{47}$$

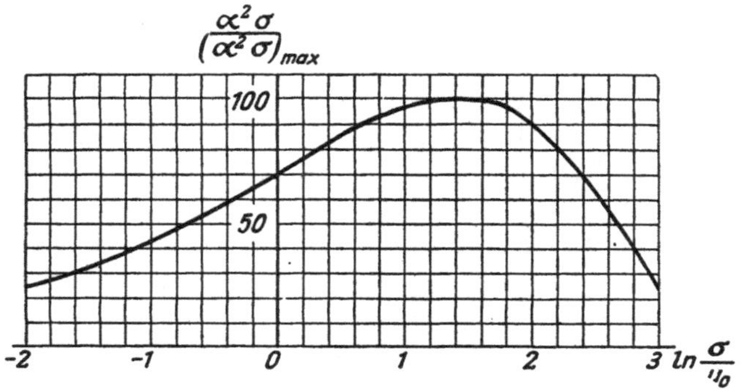

Abb. 6

Bei diesem n_{opt} liefert die Formel (21):

$$\alpha = \frac{k}{e}(1 + \ln 2{,}7) = 2\frac{k}{e}.$$

Dies ist eine Konzentration, die schon nahezu im Entartungsgebiet der Elektronen liegt.

Abbildung 7 zeigt die Ergebnisse von Messungen an p-leitendem PbSe und p- und n-leitendem PbTe mit verschiedenen Dotierungen und mit entsprechend verschiedener Ladungsträgerkonzentration.

Soweit man aus der Temperaturabhängigkeit der Beweglichkeit schließen kann — diese hat die Form $uT^3 = \text{const}$ oder $uT^{5/2} = \text{const}$ — hat man in Formel (21) bei diesen Substanzen entweder $A = 1/2$ oder $A = 1$ zu nehmen, falls man $m^*/m_0 = 1$ setzt. Die experimentellen Werte liegen bis $n = 4 \cdot 10^{18}$ gut auf den Geraden, die der Formel (21) entsprechen. Setzen wir $m^*/m_0 = 1$, so ergibt sich aus diesen Geraden $A = 1{,}2$ für p-leitendes PbTe, $A = 1$ für p-leitendes PbSe und $A = 1/2$ für n-leitendes PbTe.

Statt dessen können wir auch annehmen, daß alle diese in ihrer chemischen Struktur sehr ähnlichen Verbindungen durch die Beziehung $uT^{5/2} = \text{const}$ und $A = 1$ gekennzeichnet werden. In diesem Fall müssen die effektiven Massen verschieden sein:

$m^*/m_0 = 1{,}13$ für p-PbTe, $m^*/m_0 = 1$ für p-PbSe und $m^*/m_0 = 0{,}7$ für n-PbTe.

Im unteren Teil der Abbildung 7 sind Werte von $\alpha^2 n$ aufgetragen, die sich aus dem oberen Teil (Kurven 1 — 3) ergeben, ferner die Werte $\alpha^2\sigma$, die unter der Voraussetzung berechnet wurden, daß für alle Proben $A = 1$ ist und der Zusammenhang zwischen Beweglichkeit und effektiver Masse in diesem Fall durch die Beziehung

$u m^{*5/2}$ = const gegeben wird. (Kurven 1a und 3a, entsprechend den Kurven 1 — 3 im oberen und unteren Diagramm). Wie aus dem unteren Teil der Abbildung hervorgeht, wird das Maximum von $\alpha^2\sigma$ in allen Fällen bei $\alpha = 172\,\mu$ V/Grad erreicht; es ist am größten bei p-PbTe, für das m^*/m_0 am kleinsten ist. Der Maßstab für $\alpha^2\sigma$ in der unteren Abbildung ist so gewählt, daß für $m^*/m_0 = 1$ die Größen $\alpha^2 n$ und $\alpha^2\sigma$ zusammenfallen.

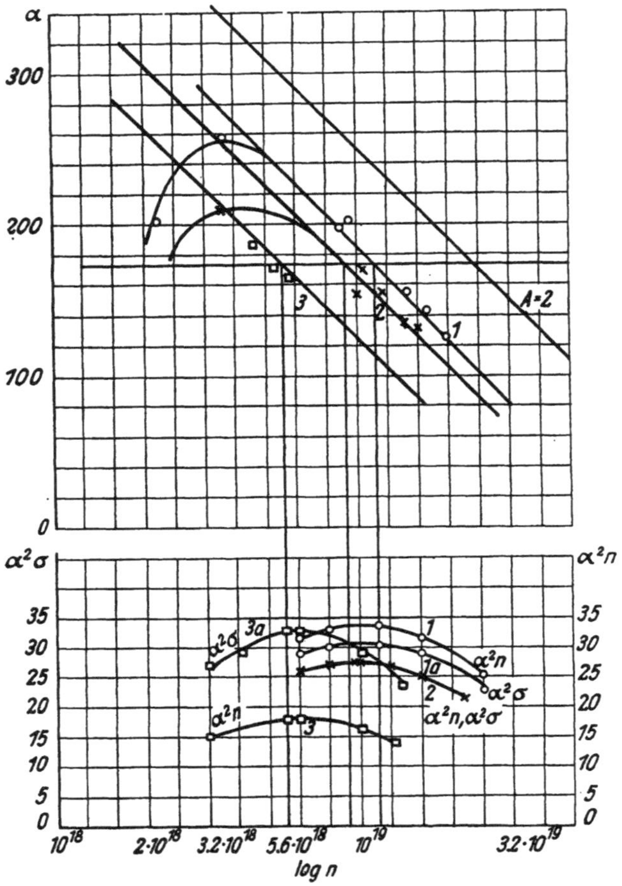

Abb. 7. Vergleich der experimentellen Daten mit Formel (21). ○ : p-leitendes PbTe ($A = 1{,}2$; $m^*/m_0 = 1$ oder $A = 1$; $m^*/m_0 = 1{,}13$); □ : n-leitendes PbTe ($A = 1/2$; $m^*/m_0 = 1$ oder $A = 1$; $m^*/m_0 = 0{,}7$); × : p-leitendes PbSe ($A = 1$; $m^*/m_0 = 1$).

Im oberen Teil der Abbildung 8 ist für Atomgitter α als Funktion von n dargestellt, wobei $m^*/m_0 = 1$ (Gerade 1) bzw. $m^*/m_0 = 1/4$ angenommen ist (Gerade 2, die offenbar n-leitendem Germanium entspricht). Im unteren Teil der Abbildung 8 ist der Verlauf von $\alpha^2 n$ (Kurven 1 und 2) für beide Fälle dargestellt, wobei für das Gitter mit $m^*/m_0 = 1/4$ (Kurve 2) der Maßstab um den Faktor 10 gestreckt ist. In Halbleitern mit geringerer effektiver Masse und entsprechend kleinerem Wert von $(\alpha^2 n)_{max}$ ist $(\alpha^2\sigma)_{max}$ größer.

Neben den Kurven 1,2 für $\alpha^2 n$ sind im unteren Teil der Abbildung 8 auch solche für $\alpha^2\sigma$ eingezeichnet (Kurven 1a und 2a). Diese Kurven haben beide den gleichen Maßstab.

Die Leitfähigkeit σ ergibt sich aus der Konzentration n durch Multiplikation mit der Beweglichkeit u, und diese hängt von m^* ab. Bei Atomgittern läßt sich der Zusammenhang zwischen Beweglichkeit und effektiver Masse durch $u m^{*5/2}$ = const darstellen. Für $m^*/m_0 = 0{,}25$ ist speziell $u/u_0 = 32$. Wie aus Abbildung 8 (unten) hervorgeht, ist $(\alpha^2\sigma)_{max}$ für ein Gitter mit $m^*/m_0 = 0{,}25$ viermal so groß wie wie für ein Gitter mit $m^*/m_0 = 1$.

In der Formel, die den Zusammenhang von α und $\alpha^2\sigma$ mit σ wiedergibt, treten die miteinander gekoppelten Größen u und m^* auf. Nach der Quantentheorie ist für

Atomgitter $u m^{*5/2} = \text{const} = u_0 m_0^{5/2}$, für polare Gitter $u m^{*3/2} = \text{const} = u_0 m_0^{3/2}$. Für polare Gitter läßt sich also α folgendermaßen schreiben:

$$\alpha = \frac{k}{e}\left[3 + \ln \frac{2e(2\pi m_0 kT)^{\frac{3}{2}}}{h^3} - \ln \frac{\sigma}{u_0}\right]. \tag{48}$$

Für Atomgitter erhält man

$$\alpha = \frac{k}{e}\left[2 + \ln \frac{2e(2\pi m_0 kT)^{\frac{3}{2}}}{h^3} - \ln \frac{\sigma}{u_0} \cdot \frac{m^*}{m_0}\right] =$$

$$= \frac{k}{e}\left[\left(2 + \ln \frac{m_0}{m^*}\right) + \ln \frac{2e(2\pi m_0 k T)^{\frac{3}{2}}}{h^3} - \ln \frac{\sigma}{u_0}\right]. \tag{49}$$

Hieraus entnehmen wir, daß für polare Gitter bei gegebenem σ weder α noch $\alpha^2\sigma$ vom Verhältnis m^*/m_0 abhängt. Für Atomgitter mit einem kleineren Wert von m^*/m_0 nimmt die Größe $(\alpha^2\sigma)_{\max}$ zu; bei $m^*/m_0 = 2{,}7$ hat die Konstante A den Wert 3 statt 2.

Abbildung 9 zeigt den Zusammenhang zwischen $\alpha^2\sigma$ und $\log n$ für Atom- und polare Gitter; (m^*/m_0 ist Parameter). Die Kurven in Abbildung 9, die zu verschiedenen Werten m^*/m_0 gehören, lassen sich zur Deckung bringen, wenn man auf der Abszissenachse nicht n, sondern $\varepsilon = n(m_0/m^*)$ aufträgt. Solche Kurven findet man in den Abbildungen 10 und 11. Um graphisch von der Abszisse $\log \varepsilon$ zu $\log n$ übergehen zu können, ist unten in Abbildung 11 ein Werte-Netz von $\log n$ für verschiedene m^*/m_0 dargestellt. Je kleiner m^* ist, desto kleiner ist die Konzentration n_{opt}, die zur Erreichung von $(\alpha^2\sigma)_{\max}$ notwendig ist.

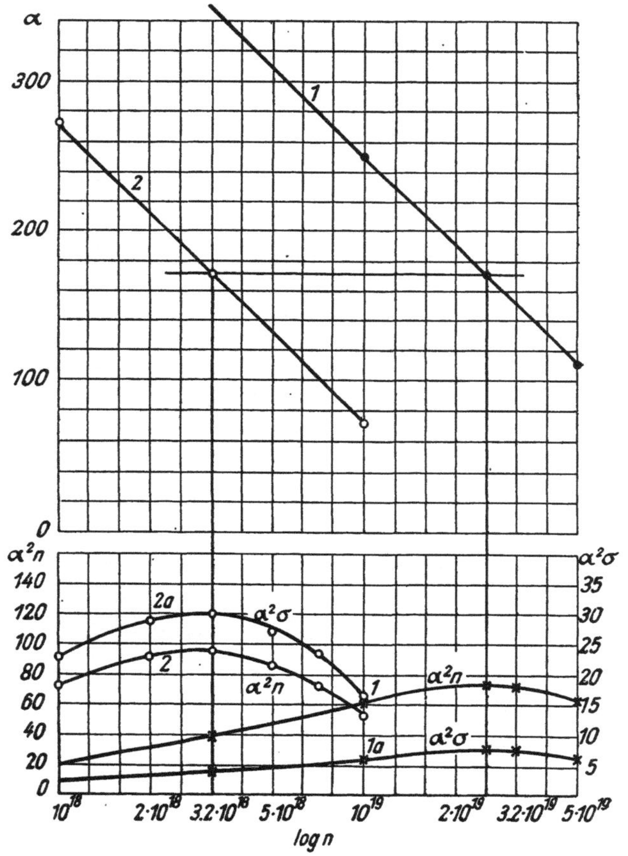

Abb. 8

Um möglichst hohe Werte von $(\alpha^2\sigma)_{\max}$ zu erzielen, wählt man also Halbleiter mit möglichst hohen Beweglichkeiten u und möglichst kleinen effektiven Massen m^*.

Wir haben den elektrischen Anteil $\alpha^2\sigma$ der Größe z betrachtet, die den erreichbaren Wirkungsgrad bestimmt. Von nicht geringerer Bedeutung sind aber die thermischen Größen des Halbleiters, die durch $\varkappa$ beschrieben werden. Hier müssen wir

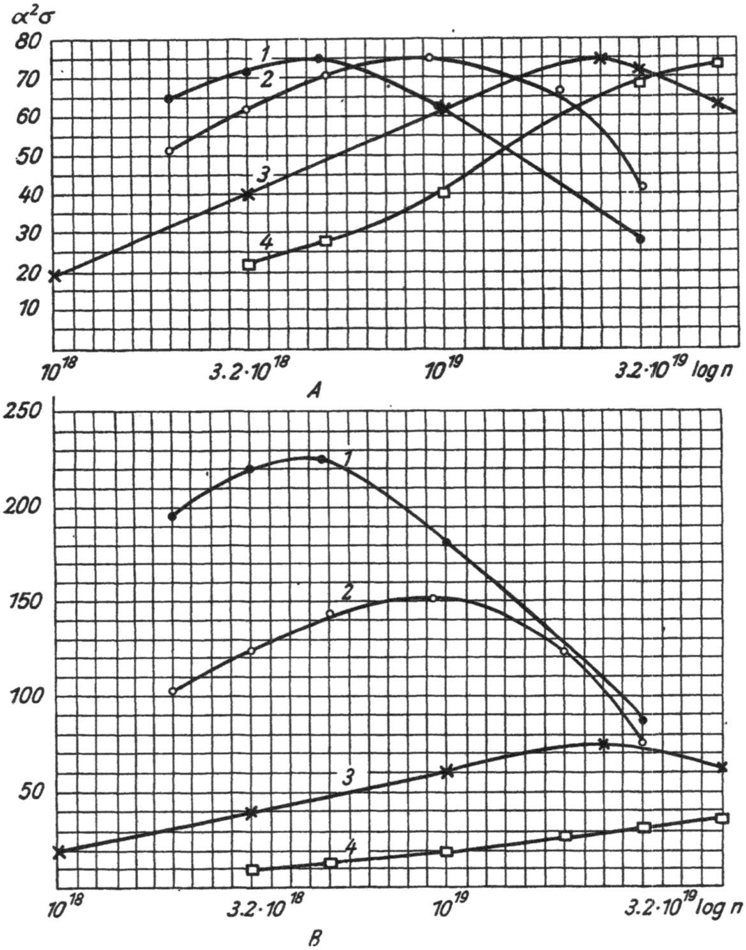

Abb. 9. A : Ionengitter, $u/u_0 = (m_0/m^*)^{3/2}$; B : Atomgitter, $u/u_0 = (m_0/m^*)^{5/2}$; 1 : $m^*/m = 1/3$; 2 : $m^*/m_0 = 1/2$; 3 : $m^*/m_0 = 1$; 4 : $m^*/m_0 = 2$.

ein Minimum anstreben. Eine geschlossene Theorie der Wärmeleitung besteht zur Zeit noch nicht. Man kann jedoch einige Regeln ableiten.

Aus Überlegungen, die aus der Vorstellung folgen, daß die Schallquantenstreuung den Wert von $\varkappa_s$ bestimmt, ferner durch einen Vergleich der in der Literatur vorhandenen Angaben mit unseren Ergebnissen kann man schließen, daß $\varkappa_s$ mit zunehmendem Atomgewicht und abnehmender Festigkeit des Materials kleiner wird. Für Ionenverbindungen ist $\varkappa_s$ kleiner als für Atomgitter.

Beispielsweise haben die Elemente und die Atomverbindungen der beiden ersten Zeilen des periodischen Systems (B, C, Si, SiC) Wärmeleitfähigkeiten von 1—2 W/Grad cm; die Valenzverbindungen und Elemente der beiden folgenden Zeilen GaSb, InSb haben Wärmeleitfähigkeiten von einigen Zehntel W/Grad cm, für Materialien der letzten Zeilen ist $\varkappa_s$ nicht höher als einige Hundertstel W/Grad cm. Eine ähnliche Abhängigkeit beobachtet man bei den Elementen der vierten Gruppe (C, Si, Ge), den Alkalihalogeniden (vom LiF bis zum RbBr), den Oxyden der Elemente der zweiten Gruppe (vom BeO bis HgO) und den Titanaten der zweiten Gruppe.

Die Wärmeleitfähigkeit jedes Stoffes läßt sich herabsetzen, wenn man im Kristallgitter Störstellen erzeugt.

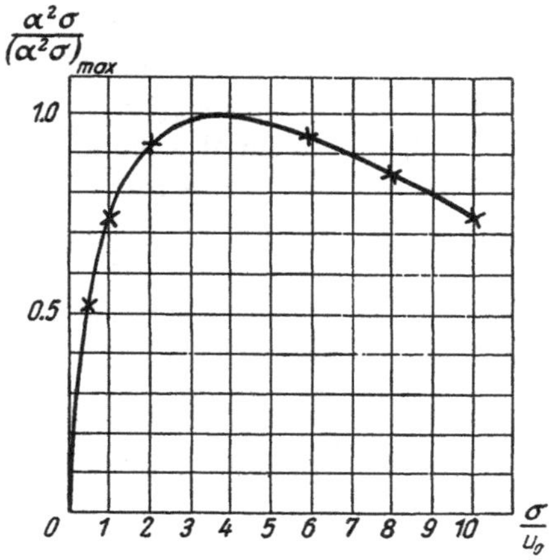

Abb. 10. Atomgitter: $u/u_0 = (m_0/m^*)^{5,2}$.

Die obigen Rechnungen gingen von der Forderung aus, $\alpha^2\sigma$ möglichst groß zu machen; sie ist mit der Bedingung, $z = \alpha^2\sigma/\varkappa$ zu einem Maximum zu machen, nur dann identisch, wenn $\varkappa$ nicht von σ abhängt. Von den beiden Anteilen, aus denen sich die Wärmeleitfähigkeit zusammensetzt

$$\varkappa = \varkappa_s + \varkappa_{el} ,$$

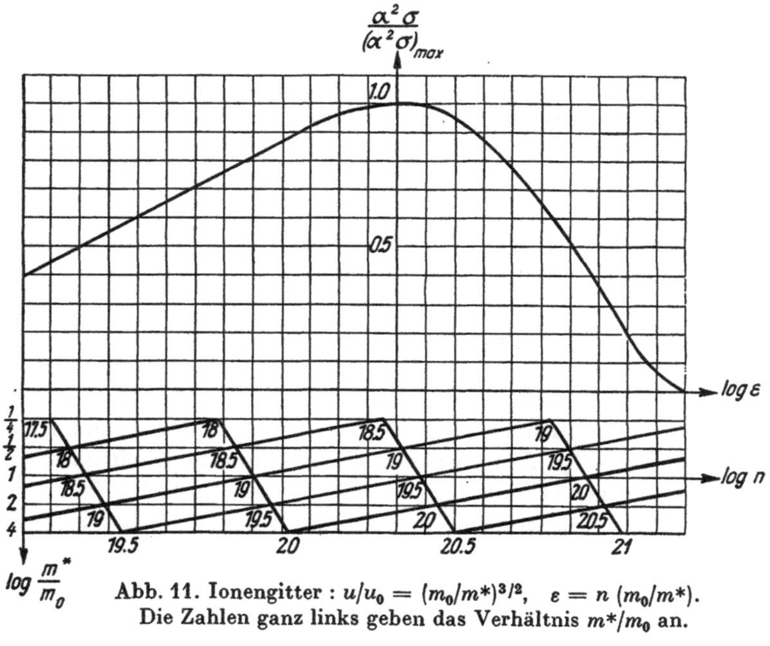

Abb. 11. Ionengitter: $u/u_0 = (m_0/m^*)^{3/2}$, $\varepsilon = n\,(m_0/m^*)$.
Die Zahlen ganz links geben das Verhältnis m^*/m_0 an.

ist nur der Schallquantenanteil unabhängig von σ, während der elektronische Anteil gerade von σ abhängt; er ist der Leitfähigkeit proportional:

$$\varkappa_{el} = L\sigma .$$

Diesen zweiten Anteil kann man vernachlässigen, wenn $\varkappa_{el} \ll \varkappa_s$ ist, was durchaus nicht immer zutrifft. Im allgemeinen Fall, wenn beide Anteile die gleiche Größenordnung haben, gilt $\alpha_{\text{opt}} = 172\ \mu$ V/Grad nicht mehr.

Statt z zu einem Maximum zu machen, stellen wir die gleichwertige Forderung, daß $1/z$ ein Minimum wird:

$$\frac{1}{z} = \frac{\varkappa_s}{\alpha^2 \sigma} + \frac{\varkappa_{el}}{\alpha^2 \sigma} = \frac{\varkappa_s}{\alpha^2 \sigma} + \frac{L}{\alpha^2} .$$

Zwischen α und σ besteht der Zusammenhang

$$\alpha = C - \frac{k}{e} \ln \sigma$$

und daraus folgt $\frac{d\sigma}{d\alpha} = -\frac{\varepsilon \sigma}{k}$.

Setzen wir $\frac{d\left(\frac{1}{z}\right)}{d\alpha} = 0$, so ergibt sich

$$\alpha_{\text{opt}} = 2\frac{k}{e}\left(1 + \frac{\varkappa_{el}}{\varkappa_s}\right) = 172\left(1 + \frac{\varkappa_{el}}{\varkappa_s}\right)\mu \text{ V/Grad}. \qquad (40\text{a})$$

Für $\varkappa_{el} \ll \varkappa_s$ geht diese Beziehung in die frühere $\alpha_{\text{opt}} = 172\ \mu$ V/Grad über.

In der Gleichung (40a) hängen sowohl die linke Seite (α) als auch die rechte ($\varkappa_{el}$) von der Konzentration n ab. Diese Gleichung ist als als transzendente Gleichung für denjenigen Wert von n aufzufassen, bei dem $\alpha^2\sigma$ ein Maximum wird.

Alle hier angegebenen Formel gelten für nichtentartetes Elektronengas. Liegt Entartung vor, so hat man die Ausdrücke (26) und (26a) zu benutzen und die Rechnungen graphisch durchzuführen.

Der für die thermoelektrischen Anwendungen sehr wichtige Zusammenhang von Wärmeleitfähigkeit und Beweglichkeit ist noch nicht gründlich genug untersucht. Man kann nur sagen, daß die Verhältnisse bei den schweren Elementen günstiger sind als bei den leichten.

Die Wärmeleitfähigkeit eines Thermoelementes ist nicht nur für den Wirkungsgrad, sondern auch zur Bestimmung der Abmessungen thermoelektrischer Batterien und für die Konstruktion der wärmetechnischen Anlagen, die den heißen Lötstellen die Wärmeenergie zuführen und sie von den kalten ableiten, von großer Bedeutung.

Damit eine Batterie bei gegebener Zusammensetzung ihren höchstmöglichen Wirkungsgrad erreicht, muß die Temperatur der heißen Lötstelle T_1 möglichst hoch, die Temperatur T_0 möglichst niedrig sein. Diese Forderungen bestimmen die Temperaturdifferenz $T_1 - T_0$. Je kleiner nun die Wärmeleitfähigkeit $\varkappa$ ist, desto kürzer kann das Thermoelement, d. h. also die Länge l, sein (Abb. 1 S. 28).

Die Dichte des Wärmestromes q (definiert als die sekundlich einen Querschnitt von 1 cm² durchsetzende Energie), der durch das Thermoelement fließt, läßt sich folgendermaßen schreiben:

$$q = \varkappa \frac{T_1 - T_0}{l} . \qquad (50)$$

Durch passende Wahl von l können wir einen bestimmten Wert von q garantieren, und umgekehrt wird bei gegebener Wärmestromdichte die Länge l der Elemente der Batterie durch die Wärmeleitfähigkeit $\varkappa$ gegeben.

In realen Thermoelementen, deren Wirkungsgrad nicht höher als 10% ist, geht der überwiegende Teil der von der Quelle gelieferten Heizleistung durch Wärmeleitung zu den kalten Lötstellen über. Die Leistung des Wärmestromes, der durch die beiden Schenkel des Thermoelementes fließt, bestimmt also in erster Linie die notwendige Leistung der Wärmequelle. Wir können die elektrische Leistung, welche die Thermo-Batterie liefern kann, durch ηQ_W beschreiben, wobei $Q_W = qS$ ist (S ist der Gesamt-Querschnitt der Batterie):

$$W = \eta q S = \eta \varkappa \frac{T_1 - T_0}{l} S.$$

Die durch 1 cm³ der Batterie erzeugte Leistung oder die spezifische Leistung beträgt

$$w = \frac{W}{S \cdot l} = \eta \varkappa (T_1 - T_0) \frac{1}{l^2}, \tag{51}$$

oder bei gegebenem Wärmestrom q:

$$w = \eta q \frac{1}{l}. \tag{51a}$$

Je kleiner die Wärmeleitfähigkeit ist, desto kürzer kann das Thermoelement sein, damit sich eine bestimmte Temperaturdifferenz einstellt; oder um so größer muß bei gegebenem W die Wärmestromdichte sein. Mit abnehmender Länge l wächst die spezifische Leistung wie l^{-2}, die Wärmestromdichte wie l^{-1}.

Schließlich ist die Wärmeleitfähigkeit auch für die Bestimmung der in der Batterie auftretenden Spannungen und Deformationen bei ungleichmäßiger Erwärmung wesentlich.

Wir bezeichnen den Temperaturkoeffizienten der linearen Ausdehnung mit p. Infolge der Temperaturdifferenz $T_1 - T_0$ zwischen den heißen und den kalten Enden ändern sich deren Linearabmessungen P, die bei der Herstellung gleich waren, um die Länge

$$\Delta P = p(T_1 - T_0)\, P.$$

Haben die Elemente die Länge l, so hat die Verbiegung, die jeder Schenkel des Thermoelements unter dem Einfluß des Temperaturgradienten $(T_1 - T_0)/l$ erfährt, den Krümmungsradius¹)

$$R = \frac{P}{\Delta P} l = \frac{1}{p\,(T_1 - T_0)} l. \tag{52}$$

Bei $p = 2 \cdot 10^{-5}$ Grad^{-1}, $T_1 - T_0 = 300°$ K und $l = 0{,}6$ cm erhalten wir beispielsweise $R = 100$ cm.

Infolge dieser Verbiegung bewegt sich eine ursprünglich ebene Oberfläche der heißen Lötstelle, die an einer Kante fest anliegt, an der gegenüberliegenden um die Strecke

$$\Delta l = \frac{P^2}{2R} = P^2 \frac{p\,(T_1 - T_0)}{2l}. \tag{52a}$$

¹) Die hier und im folgenden offenbar gemeinten Verhältnisse findet man in der von uns hinzugefügten Abbildung 11a (stark übertrieben gezeichnet). — Anm. d. Red.

Bei den obigen Werten und einem Schenkel-Querschnitt von $p^2 = 10\ \mathrm{cm}^2$ weicht die zweite Kante um $\Delta l = 0{,}5$ cm von der Ebene ab; wenn die Mittellinie der Querschnittsebene befestigt ist, verschieben sich die Kanten links und rechts davon um 2,5 mm.

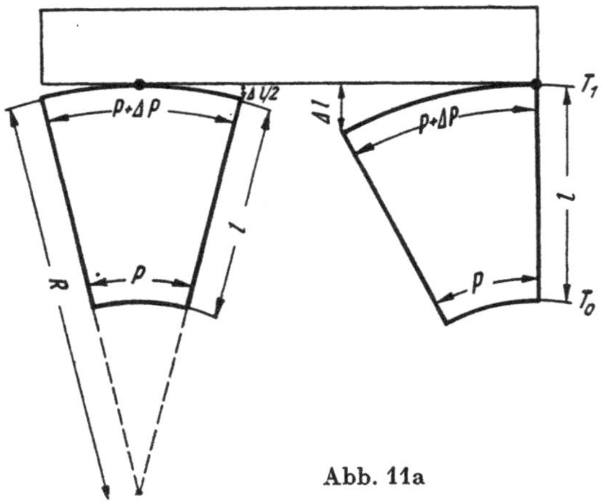

Abb. 11a

Ist aber die freie Verbiegung der Batterie erschwert, so können die in ihr auftretenden Spannungen zu ihrer Zerstörung, mindestens aber zur Bildung von Rissen führen. Bei gegebener Temperaturdifferenz $T_1 - T_0$ sind diese Spannungen um so gefährlicher, je kleiner l und je größer P ist.

Bisher haben wir angenommen, daß unser Thermoelement aus Halbleitern mit reiner Löcherleitung in dem einen Zweig und mit reiner Elektronenleitung in dem anderen besteht. Sind aber gleichzeitig freie Elektronen mit der Konzentration n_- und der Beweglichkeit u_- und Löcher mit der Konzentration n_+ und der Beweglichkeit u_+ vorhanden, so tritt an die Stelle der Formel (21) die folgende:

$$\alpha = \frac{k}{\sigma}\left\{n_- u_-\left[A + \ln\frac{2\,(2\pi m_-^* kT)^{\frac{3}{2}}}{h^3} - \ln n_-\right] - \right.$$
$$\left. - n_+ u_+\left[A + \ln\frac{2\,(2\pi m_+^* kT)^{\frac{3}{2}}}{h^3} - \ln n_+\right]\right\}. \qquad (53)$$

Bei einem chemisch reinen Halbleiter ohne Störstellen ist $n_- = n_+ = n$; wenn außerdem $m_+^* = m_-^* = m^*$ ist, bekommt man für die Thermokraft:

$$\alpha = \frac{k}{\sigma}\left\{(n_- u_- - n_+ u_+)\left[A + \ln\frac{2\,(2\pi m^* kT)^{\frac{3}{2}}}{h^3 n}\right]\right\}.$$

Für $n_- u_- = n_+ u_+$ ist $\alpha = 0$; ganz allgemein liefern Halbleiter mit gemischtem Leitungsmechanismus viel kleinere Werte von α als Störstellenhalbleiter mit Ladungsträgern einheitlichen Vorzeichens. Eine Ausnahme bilden die Halbleiter mit sehr großem Verhältnis u_-/u_+. Beispielsweise ist bei InSb dieses Verhältnis größer als 80. Damit in solchen Eigenhalbleitern hinreichend hohe Konzentrationen n auftreten, darf die verbotene Zone nicht allzu breit sein.

In unseren bisherigen Rechnungen haben wir die THOMSON-Wärme nicht berücksichtigt, die vom Strom wegen des Temperaturgradienten $(T_1 - T_0)/l$ in den beiden Schenkeln des Thermoelements erzeugt oder verbraucht wird.

Wenn die Maximalbedingung für $a^2\sigma$ in allen Teilen des Stromkreises erfüllt ist und a dort den konstanten Wert $\pm 172\,\mu$ V/Grad hat, so ist der THOMSON-Koeffizient $\tau = 0$, und folglich verschwindet auch die THOMSON-Wärme. Ist diese Bedingung nicht erfüllt und ist $\frac{\partial a}{\partial T} \neq 0$, so führt der Einfluß der THOMSON-Wärme zu einer Änderung von $a_{(1)}$ an der heißen Lötstelle; es ist zu ersetzen durch:

$$\bar{a} = \frac{a_{(1)} + a_{(0)}}{2}. \tag{54}$$

Wir müssen nun auch genauer festlegen, was unter den Materialgrößen $\varkappa$ und ϱ in Formel (30) zu verstehen ist, wenn es sich um eine variable Temperatur T handelt. Beide Größen sind nämlich Funktionen der Temperatur; in (30) hat man zu setzen

$$\varkappa\varrho = \frac{1}{T_1 - T_0}\int_{T_0}^{T_1} \varkappa\varrho\, dT.$$

Beim Übergang von $T_0 = 300°\,\mathrm{K}$ zu $T_1 = 600°\,\mathrm{K}$ ändert sich $\varkappa$ etwa um den Faktor 2, ϱ um den Faktor 4 — 5.

Unter $\varkappa$ hat man nicht den Mittelwert von $\varkappa$ zu verstehen, sondern es ist sein reziproker Wert, der mittlere Wärmewiderstand $1/\varkappa$ zu mitteln, der mit der Temperatur linear ansteigt:

$$\bar{\varkappa} = \frac{2}{\frac{1}{\varkappa_0} + \frac{1}{\varkappa_1}}. \tag{55}$$

Es wäre falsch, anzunehmen, $\bar{\varkappa}$ sei gleich dem Wert $\varkappa_1$ am heißen Ende und dies damit zu begründen, daß uns nur die Wärmeverluste durch Wärmeleitung an den heißen Lötstellen des Thermoelements interessieren. Der Temperaturgradient an diesem heißen Ende ist, wenn $\varkappa$ von der Temperatur abhängt, nämlich ungleich $(T_1 - T_0)/l$. Im stationären Fall stellt sich der Temperaturgradient in den Schenkeln des Thermoelements so ein, daß

$$\varkappa_1 \left(\frac{\partial T}{\partial x}\right)_{T=T_1} = \bar{\varkappa}\,\frac{T_1 - T_0}{l} \tag{56}$$

wird, wobei $\bar{\varkappa}$ durch den Ausdruck (55) gegeben wird.

Was den mittleren spezifischen Widerstand angeht, so müssen wir zur Festlegung des Wertes ϱ, der in den Formeln für z und η auftritt, die Temperaturabhängigkeit von ϱ und die Temperaturverteilung längs des Stabes kennen. Diese Temperaturverteilung ergibt sich als Folge der Temperaturabhängigkeit der Wärmeleitfähigkeit $\varkappa$ bzw. ihres reziproken Wertes $1/\varkappa$. Da $1/\varkappa \sim T$ ist, kann man die Temperaturverteilung längs eines zylindrischen Stabes mit folgenden Formeln beschreiben:

$$\frac{dT}{T} = C\,dx, \quad T = T_1 e^{Cx}, \quad C = \frac{1}{l} \ln \frac{T_0}{T_1}.$$

Dabei ist x der Abstand von der heißen Lötstelle.

Wenn man, wie bei den meisten Halbleitern, $\varrho(T)$ durch die Formel $\varrho T^{-n} = \text{const}$ darstellen kann, wird

$$\bar{\varrho} = \frac{\int\limits_0^l \varrho\, dx}{l} = \frac{\int\limits_{T_0}^{T_1} \varrho(T) \frac{dx}{dT} dT}{l} = \frac{\text{const.}}{Cl} \int\limits_{T_0}^{T_1} \frac{T^n}{T} dT = \frac{\text{const.}}{Cl} \int\limits_{T_0}^{T_1} T^{n-1} dT, \tag{57}$$

$$\bar{\varrho} = \frac{1}{n} \cdot \frac{\text{const}}{Cl} (T_1^n - T_0^n) = \frac{1}{n} \cdot \frac{1}{C} \cdot \frac{\varrho_1 - \varrho_0}{l} = \frac{1}{n} \ln \frac{T_1}{T_0} (\varrho_1 - \varrho_0),$$

und im allgemeinen Fall für beliebiges $\varrho(T)$:

$$\bar{\varrho} = \frac{1}{Cl} \int\limits_{T_0}^{T_1} \varrho(T) \frac{1}{T} dT = \ln \frac{T_1}{T_0} \int\limits_{T_0}^{T_1} \varrho(T) \frac{1}{T} dT. \tag{57a}$$

Die Formel (21) drückt α durch die drei Größen m^*, T und n aus. Das Experiment zeigt, daß man durch Einbau von Fremdsubstanzen in einen Halbleiter oder durch Abweichungen von der genauen stöchiometrischen Zusammensetzung des Grundgitters die Konzentration n innerhalb weiter Grenzen variieren kann, ohne dabei m^* merklich zu ändern.

Fremdsubstanzen sind nicht nur in der Lage, die Konzentration, sondern auch das Vorzeichen der Ladungsträger zu verändern. Treten die Überschuß- oder Fremdatome ionisiert ins Gitter ein, und geben sie Elektronen ans Leitungsband ab, so erzeugen sie eine Elektronenleitung. Ziehen die Fremdatome Elektronen aus dem besetzten Valenzband zu sich heran, so entsteht Löcherleitung. Beispielsweise erzeugt Bleiüberschuß in PbS Elektronenleitung, Schwefelüberschuß Löcherleitung; in beiden Fällen steigt die Ladungsträgerkonzentration. In der halbleitenden Verbindung Mg_3Sb_2 hat Magnesiumüberschuß n-Leitung, Antimonüberschuß p-Leitung zur Folge; die elektrische Leitfähigkeit wächst dabei von 10^{-9} auf 0,1 bzw. auf 1000 Ω^{-1} cm^{-1}. Sauerstoffüberschuß oder, was auf das gleiche herauskommt, Mangel an Kupfer im Kupferoxydul Cu_2O setzt die Löcher-Leitfähigkeit schlagartig von 10^{-9} bis auf 10^{-2} Ω^{-1} cm^{-1} herauf.

In allen diesen Beispielen geben Atome mit geringer Ionisationsspannung, die sich ihre Elektronen im Vakuum leicht entreißen lassen (wie Mg, Pb), diese Elektronen auch im Kristall ab und erhöhen dadurch die Konzentration der freien Elektronen. Die Atome elektronegativer Elemente wie O und S, die im Gaszustand leicht negative Ionen bilden, werden im Kristall Quellen einer Löcherleitung. Man kann annehmen, daß wir es in den erwähnten Fällen mit Gittern zu tun haben, in denen Ionenkräfte überwiegen.

Es gibt aber auch Halbleiter, in denen ein Überschuß einer beliebigen Komponente die elektrische Leitfähigkeit dadurch erhöht, daß sie Ladungsträger nur einer Sorte, die für den betreffenden Stoff charakteristisch ist, erzeugt. Hier ist etwa die halbleitende Verbindung ZnSb zu nennen, in der sowohl Zink- als auch Antimonüberschuß die Löcherleitfähigkeit erhöht; das Ladungsträgervorzeichen bleibt ungeändert. Es gibt eine ganze Anzahl von Halbleitern, bei denen wie bei ZnSb das Ladungsträgervorzeichen gleichbleibt, gleichgültig welche Fremdsubstanzen eingebaut werden.

Bei Halbleitern mit reiner Valenzbindung hängt das Vorzeichen der durch eine Fremdsubstanz erzeugten Ladungsträger von der Zahl der Valenzelektronen der Gitteratome ab. Von dieser Art sind beispielsweise Germanium und Silizium. Zu jeder chemischen Bindung zwischen zwei Atomen muß hier jedes dieser Atome ein Elektron beitragen. Tritt ein Atom anderer chemischer Natur an die Stelle eines Atoms des Grundgitters, so muß es zur Gitterbindung ebenso viele Elektronen beitragen, wie die Grundgitteratome selbst. Enthält das Fremdatom mehr Valenzelektronen, als hierzu nötig sind, so geht der Elektronenüberschuß ins Leitungsband über. Reichen die Valenzelektronen nicht aus, so werden sie aus dem Valenzband ergänzt, und dort bleibt eine entsprechende Anzahl von Löchern zurück.

In dem obigen Beispiel (Ge und Si) besitzt jedes Atom 4 Valenzelektronen, die 4 Valenzbindungen mit den Nachbaratomen absättigen. Die Elemente der 1., 2. und 3. Gruppe des periodischen Systems, die keine ausreichende Anzahl von Valenzelektronen zur Verfügung haben, rufen also eine Löcherleitung, die Elemente der 5. Gruppe eine Elektronenleitung hervor.

Wichtig ist die Feststellung, daß diese Regel nur für Substitutions-Fremdatome gilt, die an die Stelle eines Grundgitteratoms treten. Bei Fremdatomen, die in Zwischengitterplätze der Elementarzellen eingebaut werden und dort nur örtliche Verzerrungen des Grundgitters hervorrufen, hängt das Vorzeichen der resultierenden Leitfähigkeit im wesentlichen von der Elektropositivität oder Elektronegativität des Fremdatoms ab. In diesen Fällen kann man erwarten, daß die Elemente der ersten Gruppen im periodischen System eine Elektronenleitung, die Elemente der sechsten und siebenten Gruppe eine Löcherleitung erzeugen, ebenso wie bei Ionenverbindungen.

Schließlich kann ein Fremdatom sich noch in verzerrten Bereichen und in Gitterlücken festsetzen, die gewöhnlich infolge verschiedener Diffusionsgeschwindigkeit der Komponenten entstehen. In diesen Fällen wird die Elektronenbeweglichkeit gewöhnlich erhöht, und das Leitungsvorzeichen hängt von den Verhältnissen ab, die beim Einbau vorliegen.

Ein Zusatz von Fremdatomen oder eine Abweichung von der genau stöchiometrischen Zusammensetzung der Verbindung ist auch auf andere Eigenschaften des Halbleiters von Einfluß: In der Regel setzt dies die Beweglichkeit von Elektronen und Löchern und die Wärmeleitfähigkeit herab. In beiden Fällen liegt die Ursache hierfür in einer Zunahme der Konzentration der Streuzentren.

Wären die Streubedingungen für Elektronen und Schallquanten gleich, so würde sich das Verhältnis der Beweglichkeit u zum Schallquantenanteil der Wärmeleitfähigkeit durch Zusatz einer Fremdsubstanz nicht ändern. In diesem Fall wäre $\varkappa_s/u = \text{const.}$

In Wirklichkeit sind jedoch, wie wir schon andeuteten, die Streubedingungen für Elektronen und Schallquanten verschieden. Man kann also $\varkappa_s$ nicht einfach als proportional zur Beweglichkeit u annehmen. So steigt für die Elemente C (Diamant), Si, Ge, Sn die Elektronenbeweglichkeit (900, 1200, 3600, 5000), während die Wärmeleitfähigkeit abnimmt (0,45; 0,26; 0,14).

Verschwindend geringe Beimischungen mit Konzentrationen unterhalb 0,1 % verringern die Beweglichkeit in Ge schon sehr, während der Schallquantenanteil der

Wärmeleitfähigkeit nur zweimal kleiner wird, wenn man beispielsweise 3% Silizium zusetzt.

Der Zusatz von Fremdsubstanzen verschiedener Art zu Halbleitern ist das wichtigste uns zugängliche Mittel, um die Eigenschaften dieser Halbleiter ($\alpha, \sigma, \varkappa$) in der gewünschten Weise zu verändern. Man darf jedoch nicht übersehen, daß die Fremdsubstanzen nicht nur die Konzentration n der freien Ladungsträger erhöhen, sondern auch ihre Beweglichkeit u herabsetzen, so daß die Leitfähigkeit σ nicht mehr proportional zu n ansteigt, sondern weniger. Beim Zusatz von Fremdsubstanzen nimmt zwar der Elektronenanteil der Wärmeleitfähigkeit ebenso zu wie σ, aber gleichzeitig nimmt der Schallquantenanteil ab. Berücksichtigt man diese Umstände, so findet man, daß das Maximum von $\alpha^2\sigma$ bei Werten der Thermokraft in der Nähe von 172 μ V/Grad liegt.

Fremdsubstanzen in einem Halbleiter beeinflussen nicht nur die Elektronenkonzentration, die Beweglichkeit und die Wärmeleitfähigkeit bei festgehaltener Temperatur, sie ändern auch die Temperaturabhängigkeit dieser Größen.

Die durch die Fremdsubstanzen erzeugten Streuzentren setzen die Wärmeleitfähigkeit eines Halbleiters herab und bewirken ein langsameres Ansteigen des Wärmewiderstandes $1/\varkappa$ mit der Temperatur.

Diese Tatsachen werden durch Abbildung 12 veranschaulicht; hier sind Meßergebnisse für die Wärmeleitfähigkeit $\varkappa$ dargestellt, und zwar einerseits an reinem Bleitellurid (Kurve 1), andererseits an dem gleichen Material mit einer reichlichen

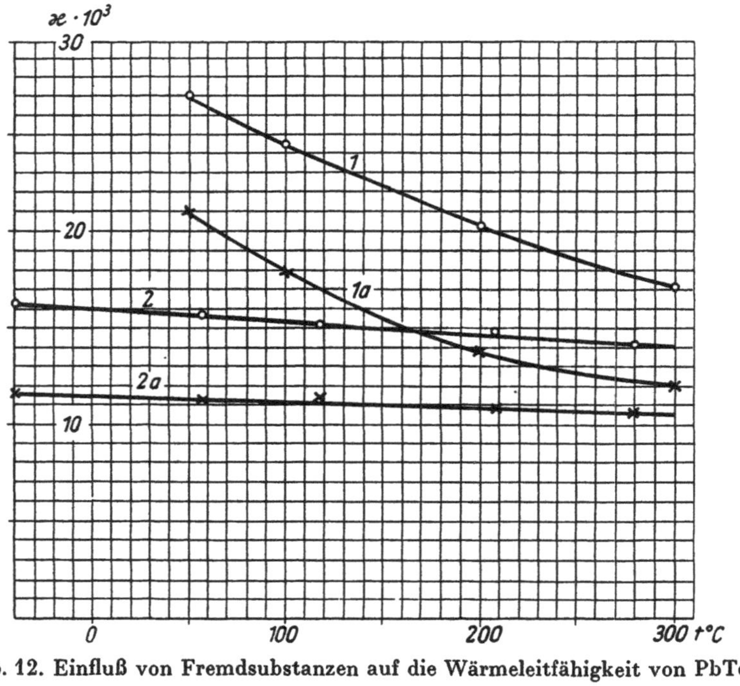

Abb. 12. Einfluß von Fremdsubstanzen auf die Wärmeleitfähigkeit von PbTe.

Dotierung (Kurve 2). Hat man die Wärme- und die elektrische Leitfähigkeit beider Substanzen in einem Temperaturintervall von —40 bis 300° C gemessen, so kann man nach dem WIEDEMANN-FRANZschen Gesetz $\varkappa_{el}$ berechnen und von $\varkappa$ abziehen und so den Schallquantenanteil der Wärmeleitfähigkeit $\varkappa_s$ erhalten (Kurven 1a und 2a).

Abbildung 12 zeigt die durch die Dotierung verursachte Verringerung des Schallquantenanteils der Wärmeleitfähigkeit. Bei 50° C hat $\varkappa_s$ auf die Hälfte abgenommen: $11{,}2 \cdot 10^{-3}$ statt $21 \cdot 10^{-3}$. Man kann auch feststellen, daß $\varkappa_s$ in der reinen Substanz annähernd umgekehrt proportional der absoluten Temperatur abnimmt:

$$\frac{\varkappa_s\,(50^\circ\,\mathrm{C})}{\varkappa_s\,(300^\circ\,\mathrm{C})} = 1{,}72,$$

während das Verhältnis der absoluten Temperaturen

$$\frac{573^\circ\,\mathrm{K}}{323^\circ\,\mathrm{K}} = 1{,}78$$

ist.

Ein ganz anderes Bild zeigt der Fremdhalbleiter: bei dem gleichen Temperaturverhältnis von 1,78 ist hier

$$\frac{\varkappa_s\,(50^\circ\,\mathrm{C})}{\varkappa_s\,(300^\circ\,\mathrm{C})} = 1{,}05.$$

Diese geringe Abnahme von $\varkappa_s$ mit wachsender Temperatur deutet darauf hin, daß die von der Temperatur unabhängige Streuung der Schallquanten an den Fremdatomen viel stärker ist als ihre Streuung an den thermischen Schwankungen[1]).

Für die Elektronenkonzentration n ist die Ionisierungsarbeit ΔE der Fremdatome in der betreffenden Substanz wesentlich, oder anders ausgedrückt, der energetische Abstand der Elektronen des Störatoms vom Rande der verbotenen Zone.

Liegt der Energieterm des Atom-Elektrons nahe dem Rande des Leitungsbandes, so führt dies zu einer Elektronenleitung; erzeugt das Atom einen Term, der näher am Rande des Valenzbandes liegt, so gehen Elektronen aus diesem Band zu den Fremdatomen über und es entsteht eine Löcherleitung. Wir bezeichnen den energetischen Abstand des Fremdatomterms von der Bandkante mit ΔE und die Anzahl der Fremdatome mit N; dann gilt[2]):

$$n_{\pm} = \frac{2^{\frac{1}{2}} N^{\frac{1}{2}} (2\pi m^* kT)^{\frac{3}{4}}}{h^{\frac{3}{2}}} \cdot e^{-\frac{\Delta E}{2kT}}.$$

Die Energie ΔE hängt von der Ionisationsspannung des Störatoms und der Dielektrizitätskonstante der Grundsubstanz ab. Für wasserstoffähnliche Atome, die in ein Medium mit der Dielektrizitätskonstante ε eingebettet sind, beträgt der Radius der ersten BOHRschen Bahn

$$r = 0{,}535 \cdot \varepsilon \cdot 10^{-8}\,\mathrm{cm}.$$

1) D. h. als die Wechselwirkung der Schallquanten mit sich selbst. — Anm. d. Red.

2) Diese Formel gilt nur unter speziellen Voraussetzungen (wie z. B. keine Entartung, geringer Ionisationsgrad der Störatome). Für allgemeinere Fälle vgl. etwa W. SHOCKLEY, Electrons and Holes in Semiconductors, New York 1950. — Anm. d. Red.

Bei einem ε von der Größenordnung 10 und darüber ist r wesentlich größer als die Gitterkonstante des Kristalls, und man kann, ohne einen großen Fehler zu machen, das Grundgitter als Kontinuum ansehen. Dann ist

$$\Delta E = E_0 \cdot \frac{1}{\varepsilon^2} = \frac{13,5}{\varepsilon^2} \text{ eV}.$$

Fremdatome auf Zwischengitterplätzen verzerren in der Regel das Gitter stärker als solche, die reguläre Gitterplätze einnehmen, namentlich das Gebiet, das den Hauptanteil der COULOMBschen Wechselwirkungsenergie des Ions mit seiner Umgebung enthält. Bei Fremdatomen auf Zwischengitterplätzen braucht also die obige Formel mit einem konstanten ε nicht immer die richtige Größe für ΔE zu liefern.

Die Temperaturabhängigkeit des elektronischen Anteils der Wärmeleitfähigkeit $\varkappa_{el}$ wird durch das WIEDEMANN-FRANZsche Gesetz gegeben:

$$\varkappa_{el} = 1.72 \cdot 10^{-8} \cdot \sigma T.$$

Wächst σ mit der Temperatur, wie das bei Halbleitern mit geringer elektrischer Leitfähigkeit meistens der Fall ist, so steigt $\varkappa_{el}$ schneller als $\sim T$ an.

In Halbleitern mit geringem spezifischen Widerstand, die man in Thermoelementen mit hohem Wirkungsgrad vorzugsweise verwendet, bleibt n in einem großen Temperaturbereich nahezu konstant, während die Beweglichkeit u mit wachsender Temperatur abnimmt. Folglich nimmt σ mit der Temperatur ab, und zwar stärker als mit T^{-1}; in diesen Substanzen nimmt deshalb $\varkappa_{el}$ mit der Temperatur ab.

In reinen Stoffen ohne Fremdsubstanz nimmt die Beweglichkeit mit wachsender Temperatur gemäß $uT^{3/2} = \text{const}$ oder $uT^{5/2} = \text{const}$ ab. Bei konstantem n nimmt folglich $\varkappa_{el}$ mit wachsender Temperatur entweder wie $T^{-1/2}$ oder wie $T^{-3/2}$ ab.

Unter Verhältnissen, wie sie in Halbleiter-Thermobatterien meistens anzuteffen sind, wenn nämlich die Elektronen durch Fremdatome erzeugt werden, können Fälle vorkommen, in denen die Fremdsubstanz die wesentliche Ursache der Streuung von Elektronen und Wärmewellen ist. Im Vergleich mit der Streuung an Fremdatomen, die unabhängig von der Temperatur ist, spielt die Streuung der Wärmewellen an den Schwankungen der thermischen Bewegung, die proportional der absoluten Temperatur ist, nur eine untergeordnete Rolle. In diesen Fremdhalbleitern ändert sich also der Schallquantenanteil der Wärmeleitfähigkeit nur wenig mit der Temperatur (vgl. Abbildung 12 Kurve 2a). Die Temperaturabhängigkeit des Elektronenanteils der Wärmeleitfähigkeit ist, wie wir bereits erwähnten, durch die Temperaturabhängigkeit des Produkts σT gegeben.

Die Streuung der Elektronen an ionisierten Fremdatomen ist um so stärker, je kleiner die Elektronengeschwindigkeit ist; in nichtentarteten Halbleitern ist die mittlere Elektronengeschwindigkeit proportional zu $\sqrt{T}$. In Halbleitern, in denen die Beweglichkeit durch die Streuung an ionisierten Fremdatomen bedingt wird (bei tiefen Temperaturen liegt dieser Fall häufig vor), nimmt sie also mit wachsender Temperatur nicht ab, sondern zu. Überwiegt die Streuung an ionisierten Fremdatomen, so hat dies noch einen weiteren wesentlichen Einfluß auf die thermoelektrischen Eigenschaften eines Halbleiters: die Abhängigkeit der freien Weglänge l von der kinetischen Energie ε hat in diesem Fall die Form

$$l \sim \varepsilon^2.$$

Das konstante Glied in der Formel (21) bekommt deshalb den Wert 4 und die Formel für Fremdleitung einheitlichen Vorzeichens lautete jetzt:

$$\alpha = \pm \frac{k}{e}\left[4 + \ln \frac{2(2\pi m^* kT)^{\frac{3}{2}}}{h^3 n}\right].$$

Für den optimalen Wert

$$\alpha = 172\left(1 + \frac{\varkappa_{el}}{\varkappa_s}\right)\mu\,\mathrm{V/Grad}$$

muß also die Elektronenkonzentration n größer sein als in dem Fall, in dem das erste Glied in der eckigen Klammer gleich 2 ist.

Der Zusatz einer Fremdsubstanz setzt zwar die Beweglichkeit herab; durch die Erhöhung der Elektronenkonzentration wird dieser Effekt mehr als ausgeglichen: bei gleichem Wert α ist also die elektrische Leitfähigkeit größer, und damit wächst auch $\alpha^2\sigma$. Da die Fremdsubstanz die Schallquanten-Wärmeleitfähigkeit $\varkappa_s$ vermindert, nimmt der Wert $z = \alpha^2\sigma/\varkappa$ bei Einführung ionisierter Fremdatome ebenfalls zu, sofern die Anzahl der Ionen so groß ist, daß sie diese wesentliche Ursache für die Streuung der Elektronen sind.

Der Einbau einer großen Zahl von Fremdatomen, die freie Ladungen eines bestimmten Vorzeichens schaffen, erhöht häufig deren Konzentration so beträchtlich, daß ein entartetes System entsteht. Mit wachsender Konzentration nimmt α ab, jedoch erfolgt im Zustand der Entartung diese Abnahme von α mit wachsendem σ langsamer als im nichtentarteten Zustand.

Der Optimalwert von α ist bei Entartung graphisch zu ermitteln, und zwar unter Benutzung von Tabellen für die Funktionen $F_r(\mu/kT)$. Bei starker Entartung $(\mu/kT \gg 2)$ kann man den Halbleiter als Metall mit geringer Ladungsträgerkonzentration und $r = 2$ auffassen. Die Minimumsbedingung

$$\frac{1}{z} = \frac{\varkappa_s}{\alpha^2\sigma} + \frac{L}{\alpha^2}$$

führt im allgemeinen Fall auf die Differentialgleichung

$$\frac{d\left(\frac{1}{\sigma}\right)}{d\alpha} = \frac{2}{\alpha}\left(\frac{1}{\sigma} + \frac{L}{\varkappa_s}\right),$$

wobei L in Abhängigkeit vom Grad der Entartung Werte zwischen $2\,\frac{k^2}{e^2}\,T$ und $3{,}3\,\frac{k^2}{e^2}\,T$ annehmen kann.

Mit einer graphischen Darstellung der Abhängigkeit $\alpha(1/\sigma)$ kann man für eine gegebene Größe $L/\varkappa_s$ den Wert α_{opt} ermitteln.

Wir haben einige Überlegungen zur Wahl von Materialien für thermoelektrische Generatoren dargelegt und untersuchen jetzt einige spezielle Fragen, wobei wir die entsprechenden Werte angeben.

Wie wir sahen, hängt der Wirkungsgrad eines Thermoelements bei optimaler Wahl der Bedingungen von der Beweglichkeit der Elektronen und der Schallquanten-

Wärmeleitfähigkeit $\varkappa_s$ innerhalb des Intervalls T_1, T_0 ab. Die gesamte spezifische Wärmeleitfähigkeit $\varkappa$ ist natürlich stets größer als ihr elektronischer Anteil $\varkappa_{el}$.

Angenommen, wir hätten die Bedingungen für das Atomgitter $\alpha = 172\ \mu$V/Grad, $\sigma \cong 4u$ verwirklicht und für $\varkappa_s$ Werte von $5 \cdot 10^{-3}$, 10^{-2} und $2 \cdot 10^{-2}$ gewählt. Ferner nehmen wir $T_0 = 300°$ K an. In den Tabellen 3 und 4 sind die diesen Bedingungen entsprechenden Werte von z und η zusammengestellt.

Wir geben weiter (in Tabelle 5) die optimalen Werte des Wirkungsgrades η sowohl bei Verwendung von Wärmequellen von geringer Temperatur $T_1 = 373°$ K als auch von sehr hoher Temperatur $T_1 = 1800°$ K an, wobei wir in allen Fällen $T_0 = 300°$ K, $\varkappa_s = 10^{-2}$ und $u = 400$ cm²/V sec annehmen.

Nachstehend folgen einige praktische Beispiele thermoelektrischer Generatoren:

1) $T_1 = 370°$ K, $T_0 = 300°$ K, $\sigma = 1600$ Ohm^{-1} cm^{-1}, $\varkappa_s = 2 \cdot 10^{-2}$, $\varkappa_{el} = 7{,}2 \cdot 10^{-3}$, $\alpha = 230\ \mu$V/Grad, $z = 3{,}1 \cdot 10^{-3}$; $\eta = 3{,}6\,\%$;

2) $T_1 = 600°$ K, $T_0 = 300°$ K, $\sigma = 1600$ Ohm^{-1} cm^{-1}, $\varkappa_s = 1{,}5 \cdot 10^{-2}$, $\varkappa_{el} = 1{,}08 \cdot 10^{-3}$, $\alpha = 295\ \mu$V/Grad, $z = 5{,}4 \cdot 10^{-3}$; $\eta = 18\,\%$.

Tabelle 3
Werte $z \cdot 10^{-3}$ in Abhängigkeit von u und $\varkappa_s$

u $\frac{\text{cm}^2}{\text{V} \cdot \text{sec}}$ \ $\varkappa_s \cdot 10^3$ $\frac{\text{Watt}}{\text{Grad cm}}$	10	20
100	1,01	0,54
200	1,78	1,0
400	2,85	1,78
800	4,1	2,8
1000	4,4	4,3
2000	5,5	5,4
10000	6,5	6,2

Tabelle 4
Werte η_{opt} (in %) bei $T_0 = 300°$ K in Abhängigkeit von der Beweglichkeit u

u \ $\varkappa_s \cdot 10^3$	10	20	10	20
	$T_1 = 600°$K		$T_1 = 1000°$K	
100	6,1	3,7	12,8	7,7
200	9,2	6,1	18,5	12,8
400	12,6	9,2	24,3	18,5
800	15,8	12,5	26,8	24,3
1 000	16,1	13,6	29,5	25,5
2 000	18,2	16,3	32,5	29,5
10 000	20,0	19,4	34,5	34,2

In beiden Fällen muß man, um $\alpha = 230$ bzw. $295\ \mu$ V/Grad bei einer elektrischen Leitfähigkeit von $\sigma = 1600\ \Omega^{-1}\ \mathrm{cm}^{-1}$ zu erreichen, sehr hohe Anforderungen an die Beweglichkeit stellen. Leider gibt es nur wenige bekannte Materialien mit einer solchen Beweglichkeit, bei denen gleichzeitig die Schallquanten-Wärmeleitfähigkeit bei Zimmertemperatur unter $2 \cdot 10^{-2}$ W/Grad cm liegt.

Tabelle 5
Werte von η_{opt} (in %) bei $T_0 = 300°$ K, $z = 2 \cdot 10^{-3}$ und $u = 400\ \mathrm{cm}^2/\mathrm{Vsec}$ in Abhängigkeit von T_1

T_1 °K	373	600	1000	1800
η_{opt}	3,55	12,6	24,3	37,5
$\eta_{\mathrm{thermod}} = \frac{T_1 - T_0}{T_1}$	19,5	50,0	70,0	83,0
$\frac{\eta_{\mathrm{opt}}}{\eta_{\mathrm{thermod}}}$	18,0	25,0	35,0	45,0

Den Eigenschaften realer Materialien angemessener sind folgende Bedingungen:

3) $T_1 = 600°$ K, $T_0 = 300°$ K, $\sigma = 1000$, $\varkappa_s = 2 \cdot 10^{-2}$, $\varkappa_{el} = 6{,}5 \cdot 10^{-3}$, $\alpha = 200\ \mu$ V/Grad, $z = 1{,}5 \cdot 10^{-3}$; $\underline{\eta = 8{,}2\%}$.

Der Optimalwert für α würde im letzten Fall $225\ \mu$ V/Grad betragen. Bei der gleichen Elektronenbeweglichkeit ist dann $\sigma = 875\ \mathrm{Ohm}^{-1}\ \mathrm{cm}^{-1}$, $\varkappa_{el} = 6{,}3 \cdot 10^{-3}$, $z = 1{,}7 \cdot 10^{-3}$; $\underline{\eta = 9\%}$.

Wir versuchen nun die Bedingungen zu formulieren, die zur Erreichung eines möglichst hohen Wirkungsgrades notwendig sind. Wie wir sahen, muß das entsprechende Material einen möglichst hohen Wert von $z = \alpha^2\sigma/\varkappa$ bis zu möglichst hohen Temperaturen T_1 der Wärmequelle besitzen.

Hierfür ist erforderlich, daß

1. der Schallquantenanteil der Wärmeleitfähigkeit der verwendeten Halbleiter möglichst klein ist. Man kann erwarten, daß dieser Bedingung Substanzen mit niedriger Debye-Temperatur, die aus schweren, schwach gebundenen Molekülen bestehen, am besten entsprechen. Der Wert von $\varkappa_s$ läßt sich durch Einbau einer Fremdsubstanz und Bildung fester Lösungen noch weiter herabsetzen, falls diese Fremdsubstanzen nicht gleichzeitig die Elektronenbeweglichkeit in gleichem Maße verringern;

2. Die Beweglichkeit der Ladungsträger (Elektronen oder Löcher) muß unter Aufrechterhaltung der ersten Bedingung möglichst hoch sein. Dabei ist es wünschenswert, wenn die Beweglichkeit durch den Einbau der Fremdsubstanz (welche die elektrische Leitfähigkeit σ erhöht) nicht zu stark herabgesetzt wird. Dieser Bedingung entsprechen am besten Atom- oder Molekülgitter mit vorwiegend akustischen Schwingungen im thermischen Spektrum. Die Beweglichkeit kann man durch

sorgfältige Temperung etwas erhöhen, evtl. auch durch Zusatz von solchen Fremdatomen, die Fehlstellen im Gitter auffüllen und dadurch dessen Störungen zum Teil beseitigen.

3. Ein Schenkel des Thermoelements muß aus einem reinen Löcherleiter, der andere aus einem reinen Elektronenleiter bestehen. Es sind hierbei Fremdhalbleiter mit kleiner Ionisierungsenergie ΔE und einer so großen Breite der verbotenen Zone zu verwenden, daß die Ladungsträgerkonzentration bis zur Temperatur T_1 praktisch allein durch die Fremdatome bestimmt wird. Nur bei einem großen Beweglichkeitsverhältnis von Elektronen und Löchern kann man auch Halbleiter mit einer schmalen verbotenen Zone benutzen.

4. Die Elektronenkonzentration muß auf der ganzen Länge des Thermoelements von der Temperatur T_1 bis T_0 so beschaffen sein, daß in allen Teilen bei den entsprechenden Temperaturen $\alpha = 172\left(1 + \frac{\varkappa_{el}}{\varkappa_s}\right) \mu$ V/Grad ist. Zu diesem Zweck gibt man den Bausteinen des Thermoelements am besten eine ortsabhängige Dotierung oder setzt sie aus mehreren Teilen zusammen. Im Bereich der tiefen Temperaturen muß die Dotierung kleiner sein als bei hohen Temperaturen, denn α hängt bei gegebenen m^* von dem Verhältnis $T^{3/2}/n$ ab.

5. Eine wichtige Voraussetzung für die praktische Eignung eines Materials ist seine Widerstandsfähigkeit gegen chemische Einwirkungen, speziell gegen Oxydation, ferner eine hinreichende mechanische Festigkeit und Elastizität, die eine Zerstörung des Thermoelements durch die auftretenden Spannungen verhütet.

6. Außer geeigneten Materialien für die beiden Schenkel des Thermoelements hat man noch ein passendes Material für die Metallbrücke zu suchen, die diese Schenkel verbindet; dieses Metall darf an der Berührungsstelle mit den Halbleitern keine zusätzlichen elektrischen Widerstände und keine zusätzlichen mechanischen Spannungen liefern.

Die letzten beiden Forderungen werden besonders wesentlich, wenn wir von einzelnen Thermoelementen zu einer Batterie aus vielen Elementen übergehen. Bei einer Thermokraft von 172 μ V/Grad in jedem Zweig und bei einer Temperaturdifferenz von 300° erzeugt jedes Element etwa 0,1 V. Man muß also beispielsweise tausend Elemente hintereinanderschalten, um eine Batterie für 100 V herzustellen. Eine elastische Verbindung zwischen den Elementen der Batterie und der Wärmequelle bzw. der Kühlung muß also die Deformationen unschädlich machen.

8. Vakuum-Thermoelemente

Die Hauptursache für die kleinen Wirkungsgrade von Halbleiter-Thermoelementen ist die Wärmeübertragung vom heißen zum kalten Ende durch Wärmeleitung. Die Streuung der Elektronen im Halbleiter und die Übertragung der auf einer freien Weglänge angesammelten Energie des Elektrons auf das Gitter schaffen im Verein mit der Wärmeübertragung durch Wärmewellen (Schallquanten) ein ungünstiges Verhältnis zwischen Beweglichkeit und Wärmeleitfähigkeit.

Unter diesen Gesichtspunkten wäre ein Thermoelement von Interesse, das auf thermischer Elektronenemission ins Vakuum beruht. Ein solcher Generator elektrischer Energie könnte aus einer Reihe von Platten bestehen, die eine hohe Temperatur T_1 haben und denen im Vakuum eine zweite Reihe paralleler Platten mit niedrigerer Temperatur T_2 gegenübersteht. Die thermische Emission nimmt so stark mit der Temperatur zu, daß man schon bei einer Temperaturdifferenz $T_1 - T_2 > 200°$ die Emission der Platten mit der Temperatur T_2 praktisch gegenüber der Emission der heißeren Platten vernachlässigen kann. Die Differenz der kinetischen Energien erzeugt in diesem System eine entsprechende Thermospannung. Die Stromstärke wird begrenzt durch die auftretende Raumladung. Diese Raumladung ist die Hauptschwierigkeit, die man bei der Entwicklung solcher Vakuum-Thermogeneratoren zu überwinden hat.

An die Stelle der Wärmeleitfähigkeit des Gitters tritt hier ein Wärmetransport über das Vakuum durch Wärmestrahlung. Berücksichtigt man aber, daß sich das thermische Emissionsvermögen fast vollständig auf 1 — 2 Atomschichten, d. h. auf eine Oberflächenschicht von etwa 10^{-7} cm Dicke beschränkt, während die Emission und Reflexion von Ultrarotstrahlung innerhalb einer Schicht von etwa $0{,}25 \cdot 10^{-4}$ cm Dicke vor sich geht, so kann man Oberflächen mit hoher Elektronenemission und geringer Lichtemission schaffen.

Man kann Reflexionskoeffizienten bis zu 95—97% erreichen, was eine entsprechende Verminderung des Emissionsvermögens bedeutet. Die schädliche Wärmeübertragung — die Hauptursache der Verluste bei Halbleiterthermoelementen — kann also bei Vakuumgeneratoren wesentlich herabgesetzt werden.

Natürlich kann ein Vakuumgenerator nur bei hohen Temperaturen $T_1 > 900°$ K arbeiten. Es hätte dagegen keinen Sinn, die Temperatur T_2 weit unter die Grenze zu senken, bei der die Elektronenemission nahezu verschwindet.

Setzt man $T_1 = 1000°$ K und $T_2 = 700°$ K, so beträgt der thermodynamische Wirkungsgrad 30%, der Wirkungsgrad der Batterie 10%. Die Temperaturdifferenz von 700 bis zu 300° K kann man zusätzlich in einem Halbleiter-Thermoelement mit einem Wirkungsgrad von etwa 10% ausnutzen, so daß der gesamte Wirkungsgrad 20% erreichen kann.

In energetischer Hinsicht ist das betrachtete Gerät günstig; mit seiner Hilfe kann man den Wirkungsgrad eines Thermoelements erheblich erhöhen. Ob man in der Praxis Wärmequellen hoher Temperatur dadurch ausnutzen kann, daß man ein Halbleiter-Thermoelement mit einem elektronischen Generator kombiniert, hängt von der spezifischen Leistung dieses Generators, seinen Abmessungen und seiner Widerstandsfähigkeit ab; alle diese Faktoren hängen weitgehend mit der erreichbaren Dichte des Emissionstroms zusammen.

KAPITEL III

Weitere Anwendungen von Thermoelementen

9. *Kälteerzeugung*

Das Problem der thermoelektrischen Kühlung wird in einer Monographie von A. F. JOFFÉ, STILBANS, STAWIZKI, und IORDANISCHWILI „Thermoelektrische Kühlung" eingehend behandelt. Wir beschränken uns hier deshalb auf die Mitteilung der wichtigsten Ergebnisse.

Die thermoelektrischen Kühlanlagen nutzen den PELTIER-Effekt aus, d. h. die Erzeugung oder den Verbrauch von Wärme an der Grenzfläche zweier Leiter, durch die ein Strom fließt (Abb. 1, S. 42). Die Richtung des Stroms I wählen wir so, daß in einem gekühlten Raum bei der Temperatur T Wärme verbraucht (aufgenommen) und bei Zimmertemperatur T_0 wieder erzeugt (abgegeben) wird. Pro Sekunde wird dann vom Thermoelement die Energie

$$Q = \alpha I T$$

verbraucht und die Energie

$$Q_0 = \alpha_0 I T_0$$

erzeugt.

Die sekundlich im Element erzeugte JOULEsche Wärme beträgt

$$W = I^2 r,$$

wobei r der Widerstand der beiden Zweige des Thermoelements ist; die Hälfte von W geht zur heißen, die andere Hälfte zur kalten Lötstelle über.

Infolge der Wärmeleitung wird von der heißen zur kalten Lötstelle sekundlich die Energie

$$Q_W = k\,(T_0 - T)$$

transportiert[1]).

Außerdem müssen wir die Arbeit des Stroms berücksichtigen, die dieser gegen die Potentialdifferenz zwischen der heißen und der kalten Lötstelle zu leisten hat:

$$Q = \alpha I (T_0 - T).$$

Den Wirkungsgrad K, den wir in diesem Fall als Kühlungskoeffizient bezeichnen, definieren wir als das Verhältnis der von der kalten Lötstelle verbrauchten Leistung zur Leistung des elektrischen Stroms:

$$K = \frac{\alpha I T - \frac{1}{2} I^2 r - k\,(T_0 - T)}{I^2 r + \alpha\,(T_0 - T)\,I}\,. \tag{58}$$

[1]) In diesem Kapitel bezeichnen wir mit k den Wärmeleitwert, mit K den Wirkungsgrad oder Kühlungskoeffizienten.

Ähnlich wie bei der Behandlung des thermoelektrischen Generators drücken wir den Wärmeleitwert k des Thermoelements und seinen Widerstand r durch die spezifischen Werte der beiden Zweige $\varkappa_1$ und $\varkappa_2$, ϱ_1 und ϱ_2 und die Querschnitte der Zweige S_1 und S_2 aus, wobei $l_1 = l_2 = l$ ist. Der Maximalwert von K liegt, wie im Fall des thermoelektrischen Generators, bei

$$\frac{S_1}{S_2} = \sqrt{\frac{\varrho_1}{\varrho_2} \cdot \frac{\varkappa_2}{\varkappa_1}}.$$

Unter diesen Verhältnissen ist

$$kr = \left(\sqrt{\varkappa_1 \varrho_1} + \sqrt{\varkappa_2 \varrho_2}\right)^2.$$

In dem Ausdruck für K multiplizieren wir Zähler und Nenner mit r und ersetzen Ir durch V. Kürzen wir dann wieder durch V, so ergibt sich

$$K = \frac{\alpha T - \frac{1}{2} V - kr (T_0 - T) \frac{1}{V}}{V + \alpha (T_0 - T)}.$$

Wir wählen die Potentialdifferenz V zwischen den Enden des Thermoelements so, daß K ein Maximum wird. Wir setzen also $\frac{\partial K}{\partial V} = 0$, was zu der Gleichung

$$V_{\text{opt}}^2 - \frac{4kr}{\alpha} \cdot \frac{T_0 - T}{T_0 + T} V_{\text{opt}} - 2kr \frac{(T_0 - T)^2}{T_0 + T} = 0$$

führt; daraus folgt $(z = \alpha^2/kr)$:

$$V_{\text{opt}} = \frac{2kr}{\alpha} \cdot \frac{T_0 - T}{T_0 + T} \left[1 + \sqrt{1 + \frac{1}{2} z (T_0 + T)}\right].$$

Wir führen wieder, wie bei der Bestimmung des Wirkungsgrades einer Thermobatterie, die Bezeichnung

$$\sqrt{1 + \frac{1}{2} z (T_0 + T)} = M \tag{58a}$$

ein; dann erhalten wir

$$V_{\text{opt}} = \alpha (T_0 - T) \frac{1}{M - 1}. \tag{59}$$

Diesem Wert von V entspricht ein Kühlungskoeffizient

$$K = \frac{T}{T_0 - T} \cdot \frac{M - \frac{T_0}{T}}{M + 1}. \tag{60}$$

Die maximale Temperaturdifferenz $(T_0 - T)_{\max}$, die sich bei gegebenem M erreichen läßt, erhalten wir nach (60), wenn wir $K = 0$ setzen. In diesem Fall ist $M = T_0/T$ und nach (58a)

$$(T_0 - T)_{\max} = \frac{1}{2} z T^2. \tag{61}$$

Weiter wird $V_{\text{opt}} = \alpha T$, und die erforderliche Leistung beträgt

$$W = \frac{\alpha^2}{r} \cdot T^2 = kzT^2.$$

In dem Ausdruck für K stellt der erste Faktor den thermodynamischen Wirkungsgrad eines reversiblen CARNOTschen Kreisprozesses dar; der zweite Faktor gibt die Verringerung des Wirkungsgrades infolge der irreversiblen Prozesse wie Wärmeleitung und Erzeugung JOULEscher Wärme an.

Aus dem Ausdruck

$$\frac{1}{K} = \left(\frac{T_0}{T} - 1\right) \frac{M + 1}{M - \frac{T_0}{T}}$$

entnimmt man leicht, daß der Wirkungsgrad K um so kleiner ($1/K$ um so größer) ist, je größer T_0/T und je kleiner bei gegebenem T_0/T der Wert von M ist, je kleiner also $1/2\, z(T_0 + T)$ ist.

Als Beispiel setzen wir $z = 2 \cdot 10^{-3}$, $T_0 = 300°\,K$ und $\Delta T = 30°\,K$; dann wird $M = 1{,}225$ und $K = 10 \cdot \frac{0{,}144}{2{,}255} = 0{,}64$.

Da die Forderungen, denen das Material eines Thermoelements genügen muß, um K möglichst groß zu machen, wie bisher auf einen möglichst großen Wert von $z(T_0 + T)$ hinauslaufen, wird die Auswahl der Materialien durch die gleichen Überlegungen wie früher bestimmt (abgesehen davon, daß der Schmelzpunkt des Materials möglichst hoch liegen sollte).

In Tabelle 6 und Abbildung 13 sind die Werte K_{opt} in Abhängigkeit von der Temperaturdifferenz $T_0 - T$ mit z als Parameter dargestellt. Neben den Werten

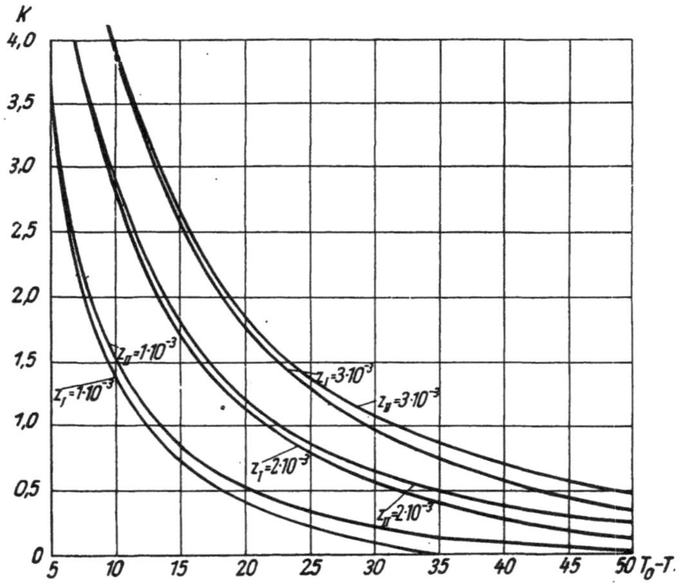

Abb. 13

$(T_0 - T)_{max}$, die sich aus (59) und (61) ergeben, enthält die Tabelle 6 auch Werte dieser Größen für eine thermoelektrische Batterie, die aus zwei unmittelbar hinter-

Tabelle 6

Kühlungskoeffizient K_{opt} und $(T_0 - T)_{max}$ in Abhängigkeit von

$z = \frac{a^2}{\varkappa \varrho}$ und $T_0 - T$ für $T_0 = 300^0$K

$z \cdot 10^3$	1,0		1,5		2,0		2,5		3,0		5	
Stufen	I	II	I	II	I	II	I	II	I	II	I	
$(T_0 - T)_{max}$ / $T_0 - T$	33	60	45	80	56	93	65	106	72	117	100	$\frac{T}{T_0 - T}$
5	3,6	3,7	4,9	5,0	6,3	6,3	7,6	7,6	8,9	8,9	12,4	59
10	1,4	1,6	2,2	2,25	2,8	2,85	3,3	3,4	4,1	4.15	6,1	29
20	0,44	0,52	0,83	0,89	1,1	1,2	1,45	1,5	1,8	1,85	2,7	14
25	0,2	0,30	0,55	0,64	0,78	0,86	1,0	1,05	1,3	1,35	2,0	11
30	0,1	0,23	0,33	0,45	0,56	0,65	0,78	0,85	0,96	1,06	1,6	9,0
35	—	0,16	0,21	0,34	0,41	0,50	0,56	0,66	0,74	0,85	1,3	7,6
40	—	0,11	0,12	0,26	0,29	0,38	0,44	0,56	0,58	0,70	1,02	6,5
45	—	0,07	—	0,21	0,20	0,31	0,33	0,44	0,45	0,58	0,84	5,7
50	—	0,03	—	0,15	0,11	0,24	0,22	0,33	0,33	0,47	0,68	5,0

einandergeschalteten Stufen mit den Temperaturdifferenzen $\left(T_0 - \frac{T_0 + T}{2}\right)$ und $\left(\frac{T_0 + T}{2} - T\right)$ besteht. Man kann zeigen, daß der Wirkungsgrad einer Kühlanlage aus zwei Stufen mit den Wirkungsgraden K_1 und K_2 durch die Beziehung

$$K = \frac{1}{\frac{1}{K_1} + \frac{1}{K_2} + \frac{1}{K_1 K_2}} \tag{62}$$

beschrieben wird. Für $K_1 = K_2$ ist also $K = \frac{K_1}{2 + \frac{1}{K_1}}$.

Eine elementare Rechnung zeigt, daß sich der Wirkungsgrad durch Benutzung zweier Stufen nur dann erheblich steigern läßt, wenn K wesentlich kleiner als Eins ist. Eine Verwendung von mehr als zwei Stufen ist praktisch zwecklos.

Was dagegen die Temperaturerniedrigung angeht, so senkt jede neue Stufe die Temperatur etwas weniger als $T^2 z/2$. Dabei wird allerdings der Wirkungsgrad nicht wesentlich verbessert. Die Temperatursenkung ist bei einem temperaturunabhängigen z proportional T^2, wobei T die Temperatur ist, die man nach der Kühlung erhält. Aufeinanderfolgende Stufen liefern also immer kleinere Temperatursenkungen. Ist beispielsweise $z = 2 \cdot 10^{-3}$ und $T_0 = 300°$ K, so erzielt die erste Stufe eine Kühlung um 59°, die zweite eine solche um weniger als 40°, die dritte um weniger als 29°, eine vierte weniger als 22° usw. Die erforderliche Leistung wächst mit der Anzahl der Stufen schnell an. Die größte Abkühlung, die man für $z = 2 \cdot 10^{-3}$ und $T = 100°$ K mit einer Stufe erreichen kann, beträgt 8,5°.

Die wichtigste Rolle für den Nutzeffekt einer Kühlanlage spielt ebenso wie beim thermoelektrischen Generator der Wärmestrom durch die Batterie. In einem Generator hängt er in erster Linie von der Wärmeleitfähigkeit ab. Die an die heißen Lötstellen von der Wärmequelle abgegebene Wärmemenge ist nur wenig größer als die von den kalten Lötstellen an die Umgebung abgegebene Wärme. In einer Kühlanlage liegen diese Verhältnisse etwas anders: zu der Leistung Q_k, die an der kalten Seite der thermoelektrischen Batterie zugeführt wird, kommt die zumeist beträchtlich höhere elektrische Energie (vom Betrage Q/K) hinzu, und an den heißen Lötstellen ist die Wärmemenge

$$Q_h = Q_k\left(1 + \frac{1}{K}\right)$$

zu entnehmen. Bei $K = 0{,}25$ wird den heißen Lötstellen eine fünfmal so große Leistung entnommen, wie den kalten zugeführt wird.

Je besser die Wärmeübertragungsbedingungen zwischen den Lötstellen des Thermoelements und dem umgebenen Medium sind, desto kleiner ist die Temperaturdifferenz gegen das Medium, die erforderlich ist, um der Oberfläche der Batterie die Leistungen Q_h/S und Q_k/S zu entnehmen bzw. zuzuführen, desto geringer ist die Temperaturdifferenz ΔT zwischen den Lötstellen des Thermoelements und desto größer ist sein Kühlungskoeffizient.

10. Heizung

Eine thermoelektrische Batterie, die eine Lötstelle abkühlt, führt zu einer Erwärmung der anderen. Im Gegensatz zur direkten Heizung durch JOULEsche Wärme dient im Thermoelement die elektrische Energie zum Transport eines Teils der Wärmeenergie von der kalten zur heißen Lötstelle.

Der Unterschied in der Berechnung einer solchen Heizanlage gegenüber einer Kühlanlage besteht darin, daß die zur kalten Lötstelle gehende Hälfte der JOULEschen Wärme, die den Kühlungskoeffizienten herabsetzt, hier zu der von der kalten Lötstelle herangeführten Wärme hinzuzuzählen ist.

Der Wirkungsgrad, der in diesem Fall als Heizungskoeffizient bezeichnet wird, ist demnach

$$K = \frac{\alpha I T + \frac{1}{2} I^2 r - k(T - T_0)}{I^2 r + \alpha I (T - T_0)} = \frac{\alpha T + \frac{1}{2} V - kr(T - T_0)\frac{1}{V}}{V + \alpha (T - T_0)} .$$

Der Heizungskoeffizient nimmt sein Maximum bei $\frac{\partial K}{\partial V} = 0$ an; daraus folgt

$$V_{\text{opt}} = \frac{\alpha (T - T_0)}{M - 1},$$

wobei M den früheren Wert (58a) hat.

Für $V = V_{\text{opt}}$ ist

$$K = \frac{T}{T - T_0}\left(1 - 2\frac{M - 1}{Tz}\right).$$

Mit $z = 2 \cdot 10^{-3}$, $T_0 = 270°$, $T - T_0 = 30°$, $M = 1{,}255$, $Tz = 0{,}6$ erhalten wir $T/(T - T_0) = 10$ und $K = 1{,}5$.

Tabelle 7 bringt Werte für den Heizkoeffizienten K in Abhängigkeit von z und $T - T_0$.

Mit zwei Stufen kann man K-Werte über 1,5 erreichen. Drei aufeinanderfolgende Stufen, von denen jede die Temperatur um $\Delta T/3$ erhöht, haben einen noch besseren Wirkungsgrad als zwei Stufen mit $\Delta T/2$. Bei einer Temperatursteigerung um 30° von 300 auf 330° K beträgt der Heizkoeffizient:

z	K_I	K_{II}	K_{III}
$1 \cdot 10^{-3}$	1,30	1,38	1,46
$1{,}5 \cdot 10^{-3}$	1,40	1,47	1,55
$2 \cdot 10^{-3}$	1,65	1,68	1,71

Aus Tabelle 7 entnimmt man, daß für Heizzwecke die Vorzüge einer thermoelektrischen Anlage für $z > 2 \cdot 10^{-3}$ wesentlich werden, besonders dann, wenn die Temperaturdifferenz in den Elementen nicht höher als 40° ist. Für $z = 5 \cdot 10^{-3}$ und $\Delta T = 40°$ ist die für die Heizung zur Verfügung stehende Wärme doppelt so groß wie die von der Batterie aufgenommene elektrische Energie. Die fehlende Hälfte der Wärmemenge wird in dem Thermoelement von der kalten Lötstelle her übertragen, wobei sich der Wärmeeinhalt von niedriger temperierten Quellen, beispielsweise eines Flusses oder einer Wasserleitung, ausnutzen läßt. Bei $\Delta T = 30°$ wird das gleiche Ergebnis schon für $z = 3 \cdot 10^{-3}$ erzielt.

Die thermoelektrische Heizung ist um so vorteilhafter, je kleiner ΔT ist. Um beispielsweise (für $z = 3 \cdot 10^{-3}$) eine bestimmte Wassermenge um 10° zu erwärmen, braucht man fünfmal weniger elektrische Energie und für eine Erwärmung um 5° achtmal weniger Energie als mit den üblichen elektrischen Heizgeräten.

Wollen wir eine kleine Temperaturdifferenz mittels thermoelektrischer Kühlung erzeugen, so brauchen wir für $z = 3 \cdot 10^{-3}$ zu einer Abkühlung um 10° viermal weniger, zu einer Abkühlung um 5° achtmal weniger elektrische Energie als mit einem üblichen Kühlgerät.

Tabelle 7

K_{opt} in Abhängigkeit von z und $T - T_0$ für $T_0 = 300°K$

$z \cdot 10^3$	1,0		1,5		2,0		2.5		3,0		4		5		10	$\frac{T}{T - T_0}$
Stufen / $T - T_0$	I	II	I	II	I	II	I	II	I	II	I	II	I	II	I	
5	4,9		5,8		7,0		8,3		9,7		11,6		13,6		21	61
10	3,1		3,5		4,0		4,6		5,2		6,2		7,3		10,6	31
15	2,1		2,3		2,7		3,1		3,7		4,4		5,0		7,3	21
20	1,8	1,8	2,1	2,1	2,3	2,3	2,6		3,0		3,6		4,0		5,6	16
25	1,6	1,65	1,75	1,8	2,0	2,05	2,3		2,5		3,1		3,5		4,7	13
30	1,3	1,38	1,4	1,47	1,65	1,68	1,8	1,8	2,15	2,15	2,5	2,5	2,8	2,8	4,0	11
40	1,0	1,25	1,2	1,32	1,45	1,5	1,6	1,7	1,8	1,8	2,1	2,1	2,3	2,3	3,2	8,5
50	1,0	1,18	1,05	1,22	1,25	1,34	1,45	1,5	1,6	1,6	1,8	1,85	2,0	2,05	2,7	7,0

11. Meßtechnik

Schon lange verwendet man Thermoelemente zur Messung von Temperaturen und Strahlungsenergien. Vor einigen Jahrzehnten war dies die einzige Anwendungsmöglichkeit für Thermoelemente.

Die Thermokraft von Halbleitern ist um mindestens eine Größenordnung höher als die von Metallen. Halbleiter haben α-Werte bis zu 1000 μ V/Grad, während Metalle nicht mehr als 20 μ V/Grad, Metallegierungen höchstens 50 μ V/Grad liefern.

Die Messung einer Temperaturdifferenz zwischen den beiden Lötstellen erfolgt in der Regel durch eine Kompensation der Thermospannung, bei der die Stromstärke auf Null abgeglichen wird. Sicherlich ermöglichen Halbleiter eine mindestens zehnmal so große Empfindlichkeit. Trotzdem haben Halbleiter noch kaum Eingang in die Praxis der Temperaturmessung gefunden.

In letzter Zeit benutzt man immer häufiger Halbleiterwiderstände, die sog. Thermistoren[1]).

Thermokreuze zum Nachweis von Strahlungsenergie werden aus Metallegierungen hergestellt. Wie sich vor kurzem herausgestellt hat, sind einige dieser Legierungen eigentlich typische Halbleiter; sie erreichen aber bei weitem nicht das Optimum hinsichtlich Thermokraft oder Wirkungsgrad. Die Empfindlichkeit eines Thermokreuzes wird durch diejenige elektrische Energie bestimmt, welche die absorbierte Strahlungsenergie erzeugt; um sehr schwache Strahlungen verstärken zu können, ist eine möglichst hohe Thermokraft Voraussetzung.

Die Bedingungen für den optimalen Wirkungsgrad von Thermokreuzen sind identisch mit denjenigen, die wir für Generatoren aufgestellt haben; von Thermokreuzen, die mit einem Verstärker arbeiten, verlangt man im wesentlichen möglichst hohe Thermokräfte. Die im Gebrauch befindlichen Geräte erfüllen bei weitem nicht die Anforderungen. Eine zweckmäßigere Auswahl von Halbleitermaterialien könnte die Empfindlichkeit von Thermokreuzen wesentlich steigern.

Weit häufiger verwendet man in der Praxis Photowiderstände und Bolometer. Außerdem liegt die Empfindlichkeit der gebräuchlichen Strahlungsempfänger schon nahe an der durch den Rauschpegel gegebenen Grenze. Die Praxis hat deshalb kein besonderes Interesse an einer Steigerung ihres Wirkungsgrades.

Die vorhandenen Halbleiter mit Schmelztemperaturen bis zu 4000° C ermöglichen die Temperaturmessung in technischen Prozessen bei sehr hohen Temperaturen oberhalb der Schmelzpunkte der Metalle. Wahrscheinlich werden Halbleiter-Thermoelemente in diesem Temperaturgebiet noch eine Rolle spielen.

Die technische Konstruktion von Halbleiter-Thermoelementen wird wahrscheinlich von den dünnen Drähtchen, aus denen die heutigen metallischen Thermoelemente bestehen, bald abgehen.

12. Schallgeneratoren

In Kapitel II erwähnten wir die auf thermischen Einwirkungen beruhenden Deformationen, die in einer Halbleiterplatte beim Durchgang des elektrischen Stro-

[1]) Widerstände definierter großer Widerstands-Temperaturabhängigkeit. — Anm. d. Red.

mes eintreten. Die Form, die die Platte annimmt, ist ein Ausschnitt aus einer Kugelfläche mit dem Radius

$$R = \frac{l}{p\,(T_1 - T_2)},$$

wobei l die Dicke der Platte, p der lineare Ausdehnungskoeffizient und $T_1 - T_2$ die Temperaturdifferenz der gegenüberliegenden Flächen ist.

Um diese Temperaturdifferenz zu bestimmen, denken wir uns durch die Platte einen Wechselstrom hindurchgeschickt, dieser möge eine Frequenz haben, bei der die Erwärmung der einen und die Abkühlung der anderen Fläche genügend Zeit haben, sich bis zur Mittelebene der Platte auszubreiten; deren Temperatur ist dann konstant (wir vernachlässigen die JOULEsche Wärme, die in unserem Fall etwa 1% der PELTIER-Wärme an den Oberflächen ausmacht).

Die Stromdichte in der Platte bezeichnen wir mit I. Dann ist die in einer Sekunde pro cm² Plattenoberfläche erzeugte PELTIER-Wärme $Q = \alpha I T$. Bei einer Frequenz von ν Hz werden in der Halbperiode $\frac{1}{2\nu}\,\alpha I T$ Joule erzeugt. Diese Energie erwärmt die eine Hälfte und kühlt die andere um ΔT ab:

$$\Delta T = \frac{1}{2\nu}\,\alpha I T\,\frac{2}{cl}. \tag{63}$$

Dabei ist c die spezifische Wärme pro cm³. Für die gesuchte Temperaturdifferenz erhalten wir:

$$T_1 - T_2 = 2\Delta T = \frac{2}{\nu}\,\alpha I T\,\frac{1}{cl} \tag{64}$$

und

$$R = \frac{N}{2}\cdot\frac{l^2 c}{p\alpha I T}. \tag{65}$$

Setzen wir beispielsweise $\nu = 50$, $l = 10^{-2}$ cm, $c = 2$, $p = 2 \cdot 10^{-5}$, $\alpha = 2 \cdot 10^{-4}$, $I = 40$ A/cm², $T = 300°$ K, so erhalten wir

$$T_1 - T_2 = 4{,}8\,°\mathrm{C} \quad \text{und} \quad R \cong 100\ \mathrm{cm}. \tag{66}$$

Wenn die Platte als kreisförmige Membran von 5 cm Durchmesser ausgebildet ist, beträgt die Amplitude der erzwungenen Schwingungen in diesem Fall 0,6 mm. Eine solche Membran kann als intensive Schallquelle mit einer Leistung von 2 W dienen. Bei 10^{-3} cm Dicke könnte die Membran Schallschwingungen bis 10 kHz erzeugen.

13. Kristallisation

An der Grenze von fester und flüssiger Phase einer bestimmten Substanz wird, ebenso wie an der Grenze zweier verschiedener Materialien, beim Durchgang eines Stromes I die PELTIER-Wärme

$$Q = \alpha I T$$

erzeugt bzw. verbraucht. Die bei der Kristallisation entstehende Wärme läßt sich nicht nur, wie üblich, durch Wärmeleitung, sondern auch in Form von PELTIER-Wärme durch einen Elektronenstrom abführen.

Wir betrachten die Kristallisationsbedingungen an einer Grenzfläche von 1 cm^2 bei der Schmelztemperatur T_0. Der Strom, dessen Dichte I [A/cm^2] betrage, möge durch gutleitende Metallblöcke zugeführt werden, deren Temperatur ebenfalls auf T_0 gehalten werde. Ferner führen wir folgende Bezeichnungen ein:

Spezifische Schmelzwärme	q	[cal/g] = 4,19 q [Joule/g] = Q [Joule/g]
Dichte der festen Phase	d	[g/cm^3]
Thermokraft zwischen fester und flüssiger Phase	α	[V/Grad]
spezifischer Widerstand der Flüssigkeit	ϱ_1	[Ω cm]
spezifischer Widerstand der festen Phase	ϱ_2	[Ω cm]
Schichtdicke der Flüssigkeit	L_1	[cm]
Schichtdicke der festen Phase	L_2	[cm]
Wachstumsgeschwindigkeit der festen Phase	$\frac{\Delta L}{\Delta t}$	[cm/sec]

An der Trennfläche wird sekundlich die Energie

$$W = \alpha I T_0 - \frac{1}{2} I^2 (\varrho_1 L_1 + \varrho_2 L_2) \tag{67}$$

erzeugt bzw. verbraucht.

Da die Temperaturen an der Grenze der beiden Phasen und an den Elektroden gleich sind, verteilt sich offenbar die JOULEsche Wärme zu gleichen Teilen auf die Elektroden und auf die Trennfläche. Das zweite Glied drückt die Energie aus, die die feste und die flüssige Phase pro Sekunde an die Trennfläche liefern, das erste Glied die vom Strom I abgeführte PELTIER-Wärme.

Die Kristallisationsgeschwindigkeit hängt folgendermaßen von der Leistung W ab:

$$\frac{\Delta L}{\Delta t} = \frac{W}{Qd} = \frac{\alpha I T_0 - \frac{1}{2} I^2 (\varrho_1 L_1 + \varrho_2 L_2)}{Qd}. \tag{68}$$

Dem Maximum der Kristallisationsgeschwindigkeit entspricht eine Stromdichte I_{opt}, die folgender Beziehung genügt:

$$\frac{d \frac{\Delta L}{\Delta t}}{dI} = 0. \tag{69}$$

Es wird also

$$I_{opt} = \frac{\alpha T_0}{\varrho_1 L_1 + \varrho_2 L_2}. \tag{70}$$

Für I_{opt} ist

$$\frac{\Delta L}{\Delta t} = \frac{1}{2} \frac{(\alpha T_0)^2}{Qd (\varrho_1 L_1 + \varrho_2 L_2)}. \tag{71}$$

Für $I \ll I_{opt}$ ist das zweite Glied in dem Ausdruck für W klein gegen das erste, und man kann in guter Näherung setzen

$$\frac{\Delta L}{\Delta t} \cong \frac{\alpha I T_0}{Qd}. \tag{72}$$

Die Kristallisationsgeschwindigkeit ist dann also proportional der Stromdichte I.

Wir stellen in Tabelle 8 die entsprechenden Werte für einige Metalle und Halbleiter zusammen. Die Schichtdicken L_1 und L_2 sind, außer bei Ge, so gewählt, daß $\varrho_1 L_1 + \varrho_2 L_2 \doteq \varrho_2$ wird; bei Ge ist $\varrho_1 L_1 + \varrho_2 L_2 = 0{,}1\, \varrho_2$.

Tabelle 8

Element	q	d	α [μV/Grad]	T_0	$\varrho_2 \cdot 10^6$	I_{opt} [A/cm²]	$\left(\frac{\Delta L}{\Delta t}\right)_{I=I_{opt}}$ [cm/sec]	$\left(\frac{\Delta L}{\Delta t}\right)_{I=100\ A/cm^2}$ [cm/sec]	[cm/h]
Zn	26	7,1	8	693	20	280	$1{,}0 \cdot 10^{-3}$	$0{,}63 \cdot 10^{-3}$	2,3
Bi	14	9,8	29—62	544	240	110	$2{,}6 \cdot 10^{-3}$	$2{,}5 \cdot 10^{-3}$	9
Cs	4	1,9	10	300	19	150	$7 \cdot 10^{-3}$	$0{,}63 \cdot 10^{-2}$	23
Ge	120	5,5	15	1250	1300	15	$0{,}5 \cdot 10^{-3}$	$0{,}4 \cdot 10^{-3}$	1,5

Die PELTIER-Wärme wird trägheitslos an der Stelle verbraucht oder erzeugt, an der die Kristallisation oder das Schmelzen vonstatten geht; der Zustrom von PELTIER-Wärme ist viel höher als der an JOULEscher Wärme (bis zu Stromdichten von mehreren 100 A/cm²). Unter diesen Bedingungen bestimmt also der Stromdurchgang den gesamten Verlauf der Kristallisation in der ganzen vom Strom durchsetzten Schicht.

Zur sorgfältigen Reinigung eines Stoffes von Fremdsubstanzen benutzt man das sog. Zonenschmelzverfahren, bei dem ein längs einer zylindrischen Probe beweglicher Ofen eine schmale Schmelzzone erzeugt, die sich langsam mit dem Ofen längs des Zylinders verschiebt. Man könnte den beweglichen Ofen durch einen Strom ersetzen. Der Strom schmilzt den festen Stab unmittelbar an der einen Seite der Schmelzzone auf; mit der gleichen Geschwindigkeit kristallisiert die Schmelze an der anderen Seite wieder. Ohne Änderung ihrer Breite verschiebt sich die Schmelzzone unter dem Einfluß einer Stromdichte I mit der Geschwindigkeit

$$v = \frac{I}{100}\left(\frac{\Delta L}{\Delta t}\right)_{I=100\ A/cm^2}, \qquad (73)$$

die wir oben berechnet haben.

Wir wollen hier die Aufzählung der Anwendungsmöglichkeiten des PELTIER-Effektes im Kristallisationsprozeß abbrechen und nur feststellen, daß die PELTIER-Wärme ein neues zusätzliches Mittel zur Steuerung dieses Prozesses unter konstanten Temperaturbedingungen ist.

ANHANG

I. Zusammenstellung der wichtigsten Formeln zur Berechnung der thermoelektrischen Eigenschaften eines Halbleiters mit entartetem Elektronengas

Im allgemeinsten Fall, der sowohl für ein nicht entartetes als auch für ein entartetes Elektronengas im Halbleiter gilt, lassen sich die Formeln zur Berechnung der thermoelektrischen Eigenschaften folgendermaßen darstellen:

1. Streuung an Gitterschwingungen konstanter Frequenz: für $r = -1$

$$\begin{aligned}
\alpha &= \pm \frac{k}{e}\left[(1 + e^{-\mu^*}) F_0(\mu^*) - \mu^*\right], \\
\sigma &= \frac{16\pi m^* e^2}{3h^3} \cdot \frac{l_0}{1 + e^{-\mu^*}}, \\
\varkappa_{el} &= \frac{16\pi m^* l_0}{3h^3} k^2 T \left[2F_1(\mu^*) - (1 + e^{-\mu^*}) F_0^2(\mu^*)\right], \\
A &= \left[2F_1(\mu^*) - (1 + e^{-\mu^*}) F_0^2(\mu^*)\right](1 + e^{-\mu^*}), \\
\frac{\varkappa_{el}}{\sigma} &= A\left(\frac{k}{e}\right)^2 T.
\end{aligned} \qquad (74)$$

Dabei ist $\mu^* = \mu/kT$ das reduzierte chemische Potential, $l = l_0/\varepsilon$ die freie Weglänge; $F_0(\mu^*)$ und $F_1(\mu^*)$ sind FERMI-Integrale (s. u.)

Aus den bekannten Ausdrücken für das WIEDEMANN-FRANZsche Gesetz erhält man folgende Grenzwerte: im klassischen Fall $A = 1$, bei vollständiger Entartung $A = \pi^2/3$.

Bei schwacher Entartung, also im Übergangsbereich, ist für A die allgemeine Formel zu verwenden.

2. Alle Streuungsarten, außer der oben betrachteten:

$$\begin{aligned}
\alpha &= \pm \frac{k}{e}\left[\frac{r+2}{r+1} \cdot \frac{F_{r+1}(\mu^*)}{F_r(\mu^*)} - \mu^*\right], \\
\sigma &= \frac{16\pi m^* e^2}{3h^3} l_0(T)(kT)^{r+1}(r+1)F_r(\mu^*), \\
\varkappa_{el} &= \frac{16\pi m^* l_0(T) k}{3h^3}(kT)^{r+2}\left[(r+3)F_{r+2}(\mu^*) - \frac{(r+2)^2}{(r+1)} \cdot \frac{F_{r+1}^2(\mu^*)}{F_r(\mu^*)}\right], \\
n &= 4\pi\left(\frac{2m^* kT}{h^2}\right)^{\frac{3}{2}} F_{\frac{1}{2}}(\mu^*),
\end{aligned} \qquad (75)$$

Dabei sind $F_r(\mu^*)$ die FERMI-Integrale

$$F_r(\mu^*) = \int_0^\infty \frac{x^r\,dx}{e^{x-\mu^*} + 1},$$

n ist die Trägerkonzentration im Leitungsband, $l = l_0(T) \cdot \varepsilon^r$ ist die freie Weglänge.

Den WIEDEMANN-FRANZschen Koeffizienten A kann man im allgemeinen Fall schreiben:

$$A = \left[\frac{(r+3)}{(r+1)} \cdot \frac{F_{r+2}(\mu^*)}{F_r(\mu^*)} - \frac{(r+2)^2}{(r+1)^2} \cdot \frac{F_{r+1}^2(\mu^*)}{F_r^2(\mu^*)}\right]. \tag{76}$$

Wie man zeigen kann, ergeben sich in den Grenzfällen folgende Werte: im klassischen Fall $A = r + 2$, bei vollständiger Entartung $A = \pi^2/3$.

Bei beginnender Entartung ist für A die allgemeine Formel zu verwenden.

Kombiniert man die oben angegebenen Ausdrücke und benutzt Tabellen der FERMI-Funktionen, wie sie im Anhang II gegeben werden, so kann man, auch für den Fall der Entartung, die einzelnen thermoelektrischen Größen oder ihre Kombinationen in Abhängigkeit von μ^* oder der Konzentration berechnen.

II. Tabellen der FERMI-Funktionen

Die FERMI-Funktionen für alle interessierenden Fälle sind hier im Intervall von $\mu^* = -4$ bis $\mu^* = 4$ angegeben. Man braucht nur diesen Bereich, weil für $\mu^* < -4$ die thermoelektrischen Eigenschaften sich nach den Formeln des nichtentarteten Falles mit einem Fehler von weniger als 1% berechnen lassen; für $\mu^* > 4$, also für ein stark entartetes Elektronengas, liefern die Formeln der Quantentheorie der Metalle einen Fehler von höchstens 1—2%.

In diesem letzten Fall kann man die FERMI-Funktionen durch folgende Näherungsausdrücke approximieren:

$$\int_0^\infty \frac{f(\varepsilon)\, d\varepsilon}{e^{\frac{\varepsilon-\mu}{kT}} + 1} = \int_0^\mu f(\varepsilon)\, d\varepsilon + \frac{\pi^2}{6}(kT)^2 \left(\frac{df(\varepsilon)}{d\varepsilon}\right)_{\varepsilon=\mu} + \frac{7\pi^4}{360}(kT)^4 \left(\frac{d^3 f}{d\varepsilon^3}\right)_{\varepsilon=\mu} + \dots \tag{77}$$

In der Tabelle werden die früheren Bezeichnungen benutzt.

μ^*	$F_0(\mu^*)$	$F_{\frac{1}{2}}(\mu^*)$	$F_1(\mu^*)$	$\frac{2}{3}F_{\frac{3}{2}}(\mu^*)$	$F_2(\mu^*)$	$F_{\frac{5}{2}}(\mu^*)$	$F_3(\mu^*)$	$F_4(\mu^*)$
−4	0,018	0,016	0,018	0,016	0,037	—	0,101	0,439
−3	0,049	0,043	0,049	0,044	0,099	0,20	0,298	1,194
−2	0,127	0,115	0,131	0,117	0,266	0,56	0,805	3,231
−1	0,313	0,290	0,339	0,307	0,705	1,43	2,160	8,721
0	0,693	0,678	0,823	0,769	1,803	3,68	5,682	23,340
1	1,313	1,396	1,806	1,774	4,312	8,55	14,390	60,981
2	2,127	2,502	3,513	3,691	9,445	17,60	34,300	153,290
3	3,049	3,977	6,096	6,902	18,870	31,90	75,730	363,690
4	4,018	5,771	9,627	11,752	34,490	—	154,300	814,640

Die Werte der Funktion $F_{5/2}(\mu^*)$ wurden durch graphische Extrapolation gewonnen.

Bei Fragen zur Produktsicherheit wenden Sie sich bitte an:
If you have any questions regarding product safety,
please contact:

Walter de Gruyter GmbH
Genthiner Straße 13
10785 Berlin
productsafety@degruyterbrill.com